Guest-edited by
Ena Lloret-Fritschi,
Selen Ercan Jenny
and David Jenny

CONSTRUCTING CHANGE
THE IMPACT OF DIGITAL FABRICATION ON SUSTAINABILITY

05 | Vol 94 | 2024

CONSTRUCTING CHANGE
THE IMPACT OF DIGITAL FABRICATION
ON SUSTAINABILITY

05/2024

About the Guest-Editors   5
Ena Lloret-Fritschi, Selen Ercan Jenny
and David Jenny

Introduction   6
A New Future of Construction
Digital Fabrication and Sustainability
Ena Lloret-Fritschi, Selen Ercan Jenny
and David Jenny

The Sustainable Lightness   14
of Digital Fabrication
Mario Carpo

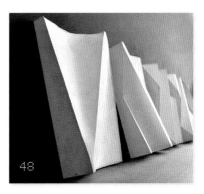

Sustainable Digital Concrete   22
Myth, Reality or Emerging Opportunity?
Timothy Wangler, Yamini Patankar
and Robert J Flatt

Injection 3D Concrete Printing   30
From Structural Geometry to Fabrication
Norman Hack, Harald Kloft, Inka Mai,
Pierluigi D'Acunto, Yinan Xiao
and Dirk Lowke

Optimising Concrete Slabs   40
with Paper Formworks
Fabio Amicarelli and Ena Lloret-Fritschi

ISSN 0003-8504   ISBN 978 1 394 23216 1

Guest-edited by **Ena Lloret-Fritschi, Selen Ercan Jenny and David Jenny**

Towards a Sustainable 3D-Concrete-Printed Architecture   48
Assemblies, Detailing and Ornamentation
Ana Anton and Benjamin Dillenburger

Reimagining Earthen Materials   56
The New Era of Sustainable and Digital Construction
David Jenny, Fabio Gramazio and Matthias Kohler

Scalable Equals Sustainable   64
The Infrastructural Imperative of Earthen Construction
Jelle Feringa

Hybrid Earth-Timber Floor Slabs   70
Scaling Circular, Low-Carbon Construction Through Automation
Tobias Bonwetsch and Tobias Huber

Look Down Not Up!   78
Solving the Building Industry's Wasteful Conundrum with Digital and Earthen Construction
Sasha Cisar

Disused Concrete, Digital Acupuncture and Reuse   88
Corentin Fivet, Stefana Parascho, Maxence Grangeot and Malena Bastien-Masse

Knowledge Production in Digital Design and Fabrication   98
Romana Rust and Inés Ariza

Rethinking Digital Construction   108
A Collaborative Future of Humans, Machines and Craft
Daniela Mitterberger and Kathrin Dörfler

Sustainable Construction Through On-site Robotic Fabrication   118
Past and Future Concepts
Selen Ercan Jenny and Abel Gawel

Concrete Patents   126
Innovative Building Materials and Construction Processes
Silke Langenberg, Sarah M Schlachetzki and Robin Rehm

From Another Perspective   134
Monitor
The Rebis Wall
Neil Spiller

There is an emerging link between digital technologies and sustainability, forming a pivotal connection within the contemporary realm of design and fabrication.
— Ena Lloret-Fritschi, Selen Ercan Jenny and David Jenny

Contributors   142

Editorial Offices
John Wiley & Sons
9600 Garsington Road
Oxford
OX4 2DQ

T +44 (0)18 6577 6868

Editor
Neil Spiller

Managing Editor
Caroline Ellerby
Caroline Ellerby Publishing

Freelance Contributing Editor
Abigail Grater

Publisher
Todd Green

Production Editor
Elizabeth Gongde

Design and Prepress
Artmedia, London

Printed in the United Kingdom by Hobbs the Printers Ltd

Front cover
Gramazio Kohler Research, ETH Zurich, Clay Rotunda, SE MusicLab, Bern, Switzerland, 2021. © Gramazio Kohler Research, ETH Zurich, SE MusicLab AG, photo Michael Lyrenmann

Inside front cover
ERNE Holzbau and Burkard Meyer Architekten, Extension of ERNE office building, Stein, Switzerland, 2023. © ERNE AG Holzbau, Stein (CH) | Bernhard Strauss, Freiburg (DE)

Page 1
Benjamin Dillenburger and Michael Hansmeyer, Tor Alva, Mulegns, Switzerland, 2024. © Michael Hansmeyer, Digital Building Technologies, ETH Zurich

EDITORIAL BOARD

Denise Bratton
Paul Brislin
Mark Burry
Helen Castle
Nigel Coates
Peter Cook
Kate Goodwin
Edwin Heathcote
Brian McGrath
Jayne Merkel
Peter Murray
Mark Robbins
Deborah Saunt
Patrik Schumacher
Jill Stoner
Ken Yeang

ARCHITECTURAL DESIGN

September/October 2024
Volume 94
Issue 05

Disclaimer
The Publisher and Editors cannot be held responsible for errors or any consequences arising from the use of information contained in this journal; the views and opinions expressed do not necessarily reflect those of the Publisher and Editors, neither does the publication of advertisements constitute any endorsement by the Publisher and Editors of the products advertised.

Journal Customer Services
For ordering information, claims and any enquiry concerning your journal subscription please go to www.wileycustomerhelp.com/ask or contact your nearest office.

Americas
E: cs-journals@wiley.com
T: +1 877 762 2974

Europe, Middle East and Africa
E: cs-journals@wiley.com
T: +44 (0)1865 778 315

Asia Pacific
E: cs-journals@wiley.com
T: +65 6511 8000

Japan
(for Japanese-speaking support)
E: cs-japan@wiley.com
T: +65 6511 8010

Visit our Online Customer Help available in 7 languages at www.wileycustomerhelp.com/ask

Print ISSN: 0003-8504
Online ISSN: 1554-2769

All prices are subject to change without notice.

Identification Statement
Periodicals Postage paid at Rahway, NJ 07065. Air freight and mailing in the USA by Mercury Media Processing, 1850 Elizabeth Avenue, Suite C, Rahway, NJ 07065, USA.

USA Postmaster
Please send address changes to Architectural Design, John Wiley & Sons Inc., c/o The Sheridan Press, PO Box 465, Hanover, PA 17331, USA

Rights and Permissions
Requests to the Publisher should be addressed to:
Permissions Department
John Wiley & Sons Ltd
The Atrium
Southern Gate
Chichester
West Sussex PO19 8SQ
UK

F: +44 (0)1243 770 620
E: Permissions@wiley.com

All Rights Reserved. No part of this publication may be reproduced, stored in a retrieval system or transmitted in any form or by any means, electronic, mechanical, photocopying, recording, scanning or otherwise, except under the terms of the Copyright, Designs and Patents Act 1988 or under the terms of a licence issued by the Copyright Licensing Agency Ltd, 5th Floor, Shackleton House, Battle Bridge Lane, London SE1 2HX, without the permission in writing of the Publisher.

𝐷 is published bimonthly and is available to purchase as individual volumes at the following prices.

Individual copies:
£29.99 / US$45.00
Mailing fees for print may apply

ABOUT THE GUEST-EDITORS

ENA LLORET-FRITSCHI
SELEN ERCAN JENNY
DAVID JENNY

Over the past decade, Ena Lloret-Fritschi, Selen Ercan Jenny and David Jenny have fostered a collaborative and innovative exploration into digital fabrication within the pioneering Chair of Architecture and Digital Fabrication at ETH Zurich. Their current research and teaching activities, at the intersection of digital technologies and architecture, span different universities in Switzerland. In both their individual and collaborative work, the overarching question driving their research revolves around understanding the impact of digital technology on architectural practices, with a focus on tectonics and aesthetics in relation to digital control over construction materials. As today's discourse has evolved towards considerations of circularity, material depletion and global warming, sustainability has emerged as a topic alongside digital technologies and has become the most significant driver behind their research endeavours.

Ena Lloret-Fritschi currently leads the professorship on Fabrication and Material Aware Architecture (FMAA) at the Academy of Architecture, Università della Svizzera Italiana (USI) in Mendrisio, which focuses on reshaping concrete and earthen materials in the built environment. Her decade-long exploration of digital concrete at ETH Zurich included the ground-breaking project Smart Dynamic Casting (SDC), which sparked her ongoing research on minimal formwork for material-optimised building elements. She guides PhD and research projects in digital technology and malleable construction materials with the goal of building with less. Her innovative work is seen in interdisciplinary patents like SDC, Eggshell and Foldcast, showcasing novel formworks and their efficiency in shaping concrete structures. Prior to her academic career, she gained experience in architectural practice before transitioning to digital fabrication during her studies at the Architectural Association (AA) in London. In 2016 she completed her PhD on the SDC robotic slipforming technique at ETH Zurich.

With a background in architecture, Selen Ercan Jenny's research has revolved around on-site robotic fabrication, starting in 2011 at the Chair of Architecture and Digital Fabrication at ETH Zurich. In 2013 she joined the Future Cities Lab at the Singapore-ETH Centre as part of the Design of Robotic Fabricated High Rises research module, exploiting the commercial potentials of applied research in construction robotics. Over the course of 10 years she has been researching and teaching digital fabrication processes, taking part in and leading interdisciplinary projects within the fields of architecture and robotics. In 2022 she co-initiated the 'Future of Construction' symposium series. She received her PhD from ETH Zurich in 2023, where she focused on developing a thin-layer, spray-based plaster printing technique eliminating additional formwork or support structures or any subtractive steps, to minimise waste. She is currently a postdoctoral researcher at ETH Zurich and at FMAA.

David Jenny is a practising architect and researcher with a background in digital fabrication and computational design methods. He is currently a senior research associate and lecturer at the Institute for Building Technology and Process at the Zurich University of Applied Sciences (ZHAW), where he is co-leading the research group Digital Technologies in Design and Fabrication. The group's focus is the development of sustainable digital constructive systems and the use of alternative building materials. After graduating from ETH Zurich in 2015, where his diploma project on algorithmic methods for housing design was awarded the Swiss Society of Engineers and Architects (SIA) master prize, he joined the Chair of Architecture and Digital Fabrication. There he was responsible for teaching digital fabrication to multiple generations of architecture students and leading the teaching projects of the Master of Advanced Studies in Architecture and Digital Fabrication postgraduate programme. ⌂

Text © 2024 John Wiley & Sons Ltd. Images: (t) © Ena Lloret-Fritschi; (c) © David Jenny; (b) © Michael Lyrenmann

A New Future

INTRODUCTION

ENA LLORET-FRITSCHI,
SELEN ERCAN JENNY
AND DAVID JENNY

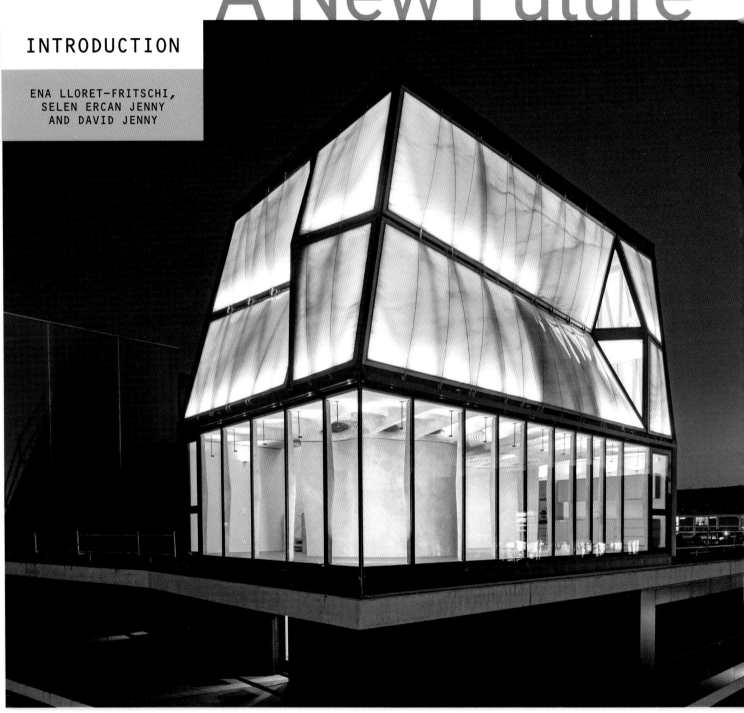

# of Construction
Digital Fabrication and Sustainability

There is an emerging link between digital technologies and sustainability, forming a pivotal connection within the contemporary realm of design and fabrication. As the urgency to move towards a sustainable building culture has become central to the disciplines of architecture and construction, this relation deserves increasing attention and significance in the discourse of today's practices. With this issue of ⌂, the focus is on projects that not only underscore an intrinsic connection between sustainability and digitisation but also actively embrace design through the lens of data, technology, fabrication and innovative construction processes.

The issue's fundamental theme revolves around showcasing the profound impact of technology in conjunction with construction materials, as a dynamic interaction with the potential to reshape the landscape of both contemporary and future architectural practice. By examining the intricate interplay between sustainability, digital fabrication, material systems and technological advancements, it aims to illuminate their transformative possibilities.

Exploring three foundational themes – concrete, earth construction and digital technologies – this issue brings together authors primarily belonging to the inaugural generation of digital fabrication researchers, architects and engineers. Firstly, Mario Carpo, the renowned architectural historian and critic, discusses and speculates on the sustainable lightness of digital fabrication and illustrates how digital technologies have shaped our production and material practices in the past – and how they might help us in tackling the big challenges of our future. Our post-industrial logic of mass-customisation allows the serial reproduction of non-identical items and thus a shift from the global transport of materials and goods towards bespoke, on-site, on-time and on-demand fabrication. Furthermore, Carpo addresses how digital tools enable us to embrace natural building materials and their circular usage thanks to intelligent robotic labour, fostering a more environmentally sustainable approach to building practice.

Perspectives on Concrete Construction

In regard to concrete, four articles highlight how this material can be reconsidered as sustainable in the future. Firstly, Timothy Wangler, Yamini Patankar and Robert J Flatt, from the Physical Chemistry of Building Materials group at ETH Zurich, discuss why, how and when concrete is sustainable, and emphasise its enduring importance as a building material. Acknowledging the ecological significance of concrete and examining its negative environmental image, they address productivity challenges, and their narrative questions the sustainability of current digital fabrication practices, providing a nuanced perspective. Through compelling examples like the DFAB House (2018) and the HiLo project (2022) in Switzerland, the authors envision a future where digital fabrication enhances sustainability by fostering creative, materially efficient designs, ultimately challenging conventional norms in construction.

```
National Centre of Competence in Research (NCCR)
Digital Fabrication,
DFAB House,
NEST, Empa Campus,
Dübendorf, Switzerland,
2019
```

The DFAB House is a collaborative demonstration module from the Swiss research consortium the National Centre of Competence in Research (NCCR) Digital Fabrication, built on the NEST (2016) – the modular research and innovation building of the Empa and Eawag research institutes where new construction technologies can be tested under realistic conditions. Researchers from eight ETH Zurich professorships cooperated with industry experts and planning professionals in a unique way to explore and test how digital fabrication can change the way we design and build.

Fabrication and Material Aware Architecture (FMAA) group,
Hochhaus Stoff,
'Soft Formwork' Spring Semester elective course,
Academy of Architecture,
Università della Svizzera Italiana (USI),
Mendrisio, Switzerland,
2023

right: Relating to the Foldcast project's research and development process, FMAA's teaching activities prioritise leveraging cutting-edge digital fabrication processes. The emphasis lies in optimising building structures through the use of recyclable ultra-thin materials such as fabric formworks. Tutors: Ena Lloret-Fritschi, Fabio Amicarelli, Selen Ercan Jenny and Yen-Fen Chan. Students: Marlene Fischer, Antonia Herting and Tobias Quezado Deckker.

Fabrication and Material Aware Architecture (FMAA) group,
Column Forest,
'Earth Building' Fall Semester elective course,
Academy of Architecture,
Università della Svizzera Italiana (USI),
Mendrisio, Switzerland,
2023

opposite: How can we build with earth in the global North using digital technologies? And how could such an architecture look today? The ongoing teaching efforts at FMAA put forward a unique challenge for students in rethinking the use of earth-based materials in architecture within the Swiss climate: getting ready to interpret brick as a new type of 3D-printed formwork used in conjunction with two infill materials – earth and concrete. Tutors: Ena Lloret-Fritschi, Sacha Cutajar and Ping-Hsun Tsai. Students: Kim Gubbini, Dea Trapaidze, Jonny Wahlfort, Jasmin Mohammadi and Agata Kobus.

Next, Norman Hack and his co-authors from the technical universities of Braunschweig (TU Braunschweig) and Munich (TUM) in Germany navigate the century-long evolution of concrete, emphasising its perpetual reinvention by architects and engineers. The advent of injection 3D concrete printing (I3DCP) signifies a transformative leap in concrete's narrative and aligns with the elegant principles of graphic statics. It employs a robotic system to inject concrete into a non-curing, non-Newtonian liquid, enabling printing along unrestricted spatial trajectories and providing unprecedented design freedoms. In the presented work, this directly translates into sustainable design considerations that explore the structural potential of graphic statics for optimal stress distribution. Moving further, Fabio Amicarelli and Ena Lloret-Fritschi underscore the imperative to minimise concrete and formwork materials, presenting their project Foldcast begun in 2022 at the Università della Svizzera Italiana Academy of Architecture in Mendrisio, Switzerland. This innovative approach integrates computational tools and digital machines to craft sustainable and structurally optimised concrete elements. The formwork is literally minimised, employing single sheets of paper. As such, Foldcast substantially reduces the construction's carbon footprint while elevating design flexibility, envisioning a paradigm shift in construction practices.

To conclude this first section, Ana Anton and Benjamin Dillenburger, from the Chair for Digital Building Technologies at ETH Zurich, present the transformative potential of 3D concrete printing (3DCP) in construction, exploring the impact of this over-80-year-old innovation by developing Tor Alva (2021–29), a tower in the village of Mulegns in the Swiss Alps designed for performance and arts as part of the Origen Foundation project, which supports different temporary architectural structures throughout the area. In the article, the authors discuss 3DCP's potential for resource efficiency, geometric complexity and material customisation, shedding light on challenges and comparing on-site versus prefabrication strategies. The Tor Alva project serves as a showcase for 3DCP's promise in terms of mass customisation and sustainability. It also addresses the planned disassembly of the tower in five years – a highly relevant consideration in today's sustainability discourse, particularly regarding the end-of-life phase of a building.

Gramazio Kohler Architects,
Gantenbein Winery,
Fläsch, Switzerland,
2006

right: In the Gantenbein project, a robotic brick-laying process was used for the construction of a winery façade, where each one of the 20,000 bricks is placed precisely according to programmed parameters at its desired angle and at the exact prescribed intervals. This early work and Gramazio Kohler Research's related Augmented Bricklaying project for the Kitrvs winery in Kitros, Greece (2019) show how technology is not time- or context-independent but reacts and becomes meaningful when integrated into specific local conditions and processes.

Earthen Construction: The Past, Present and Future

The issue delves into the topic of earth construction with four articles. Firstly, David Jenny, Fabio Gramazio and Matthias Kohler discuss how the introduction of digital technologies enables novel construction approaches with the abundant and inherently sustainable material of unfired clay. They present the Clay Rotunda (2020–21) from ETH Zurich, a project demonstrating how complex material behaviour in combination with digital technologies reshapes the way we build, moving from rigid control over industrially produced parts to a more adaptive building process that makes use of inhomogeneous materials, and how this shift paves the way for a novel aesthetic that bridges the natural world and digital innovation. Next, Jelle Feringa, a co-founder of multiple robotic fabrication companies, addresses the need for an industrial and scalable approach to earthen constructions. He discusses the paradox of earthen construction – accessible and part of human heritage across cultures and geographies – and its exclusivity today, due to the practical and economic constraints of the current artisanal building practice. He introduces the robotic shot-earth 3D printing (SE3DP) process of his startup company Terrestrial, and how this construction technique can be related to an accelerated rock cycle, where pressure is used not in geological steps but momentarily, rapidly compacting earth, gravel and clay into architectural structures. Moving on, Tobias Bonwetsch and Tobias Huber present the sustainable earth and timber floor-slab construction from their young construction startup company Rematter, which focuses on creating circular, low-carbon and equitable buildings. The innovative slab construction is developed in collaboration with engineers ZPF Ingenieure, architects Herzog & de Meuron and client Senn Resources, and is being implemented in the House of Research, Technology, Utopia and Sustainability (HORTUS) office building in Basel, Switzerland (due for completion in 2025). Rematter's industrialised digital prefabrication process promotes circular, low-carbon construction through automation, where the aim is to offer a true alternative to traditional and established construction systems. Lastly, Sasha Cisar, from radicant, Switzerland's first digital sustainability bank, discusses the means of marrying modern technology with ancient and low-carbon materials – building with local excavation material or earth and combining current construction techniques with digital fabrication and prefabrication to help the construction sector become part of the solution.

ZHAW Centre of Building Technology and Process and Università della Svizzera Italiana (USI) professorship on Fabrication and Material Aware Architecture (FMAA) in collaboration with Oxara, 'earth to earth' Winter School, USI, Mendrisio, 2023

below left: The 'earth to earth' Winter School is a real-world design and building project exploring the combination of two cutting-edge fabrication processes for earthen materials. A robotic 3D-printing process for clay is used to produce ultra-thin formwork elements that are filled with castable earth to build a fully recyclable, climate-neutral, functional staircase. Project team: David Jenny, Ena Lloret-Fritschi, Yen-Fen Chan, Sacha Cutajar, Fabio Amicarelli, Selen Ercan Jenny, Samuel Rebelo Garcia and Patric Fischli-Boson.

below right: The staircase serves as both an experimental setup to observe long-term behaviour of the proposed construction system when exposed to weather and wind, and as a proposal for the disruptive change needed in our built environment. Eventually, it dissolves back into nature.

Circularity, Data and Digital Machines

An additional theme the issue brings forward is the intricate interplay between sustainability, digital fabrication, repurposed materials and technological advancements, specifically focusing on concrete. This topic harbours diverse opinions, especially concerning the reusing of materials and building elements. Shipping building parts over large distances is associated with high energy consumption. However, the cost incurred for transportation may not outweigh the energy expended in processing building components from scratch, considering that the material for new building components must be excavated, crushed, fired, crushed again, reprocessed and transported. Here, a method to reuse what we have already broken might indeed be a sustainable approach, when considered alongside the notion that the more we excavate, the more we deplete the Earth's resources.

Corentin Fivet and his co-authors from the Swiss Federal Institute of Technology in Lausanne (EPFL) discuss the potential for waste-based circular construction and contextualise the topic within today's practice, going beyond the hype of reusing building components. They discuss how reusing load-bearing structural components demands new design processes and introduces new stakeholder dynamics, and how computational matching algorithms play a crucial role by providing fast estimates and supporting early design decisions. The presented project on digital acupuncture takes advantage of the idea of digital augmentation, where a series of scanning processes is used to digitise the geometry of large pieces of concrete rubble obtained from demolitions and transformed into new and robust load-bearing structures through the precision of robotic assembly.

Reflecting on the accelerating environmental damage caused by humans in just the last 200 years, it has not only involved colossal greenhouse gas emissions and waste, but also encompassed biodiversity loss and the disruption of ecosystems. It is imperative for us to cease recklessly depleting our planet and instead to carefully evaluate our resources and develop methods for conservation and circularity. But this requires a rethinking of processes – how we design and how we produce – and a novel understanding of how data can help us to do so. Romana Rust and Inés Ariza, through their work at Gramazio Kohler Research at ETH Zurich, discuss the rise of a digital building culture where descriptive building drawings have been replaced by prescriptive rules to build, measure and adapt, leaving behind a conglomerate of software and data repositories after their realisation that raise questions regarding their

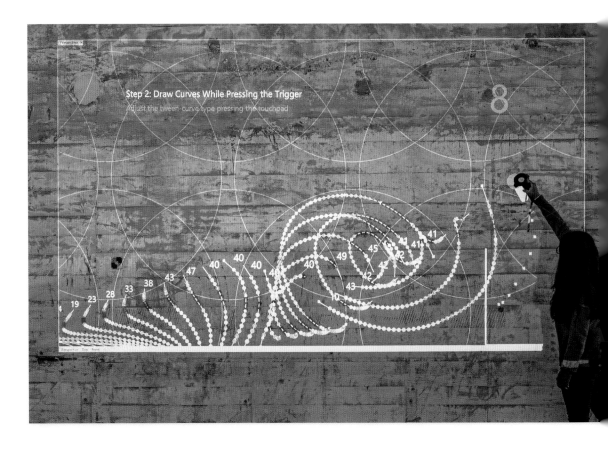

sustainability. They highlight how today the discourse on sustainability is strongly tied to material flows, performance optimisation and minimisation of material footprints. However, for these to become ubiquitous, we must fundamentally rethink how we produce, share and apply disciplinary architectural knowledge – where the concept of software as a repository for collective knowledge to be preserved, reused, reduced and recycled becomes increasingly critical.

But will digital technologies and robots take over everything? Daniela Mitterberger and Kathrin Dörfler, with their joint work from ETH Zurich and the Technical University of Munich, discuss technology's potential to empower and extend human skills to foster a meaningful collaboration between humans, machines and computational intelligence. They highlight a beautifully crafted brick façade of the Kitrvs winery in Kitros, Greece (2019), serving as a pioneering example for emphasising the indispensable role of humans, their intuition and craftsmanship, even in conjunction with digital technology. With this approach, the project revisits the initial bespoke, robotically built brick façade of the Gantenbein Winery (2006) by Gramazio Kohler Research, reintroducing craftspeople into a digital fabrication process.

Today, extended reality (XR) shapes a new approach, uniting skilled craftspeople with digital augmentation to create bespoke structures and combine human dexterity with robotic precision on construction sites – an evolution that showcases the promise of a harmonious collaboration between human ingenuity and technological advancement in the building industry. Selen Ercan Jenny and Abel Gawel discuss and speculate on the progress and challenges of on-site fabrication with their projects from ETH Zurich and the Singapore-ETH Centre, presenting a motivation to align research and development scope with sustainability goals. In addition to taking urgent action to combat climate change, these goals include ensuring healthy lives and wellbeing for all regardless of age, promoting full and productive employment and decent work for all, and fostering innovation for building resilient infrastructure. The authors look back to four research projects spanning over 10 years, introducing mobile robotic platforms on site to explore how autonomous machines could help to meet the needs of future generations.

Gramazio Kohler Research in collaboration with the groups of Robert J Flatt, Hans J Hermann and Peter Fischer,
Smart Dynamic Casting,
ETH Zurich,
2012–15

left: Smart Dynamic Casting (SDC) revisits the traditional slipforming technique and introduces a groundbreaking new digital fabrication process specifically aimed at removing the need for individually made formwork for the construction of complex concrete structures. The process was invented in 2012, and the first bespoke columns were produced at a height of up to 1.9 metres (6 feet 3 inches). Project team: Ena Lloret Kristensen (project lead and PhD), Ena Lloret-Fritschi (PhD), Andreas Thoma, Ralph Bärtschi, Thomas Cadalbert, Beat Lüdi, Orkun Kasap and Maryam Tayebani.

Gramazio Kohler Research in collaboration with the Master of Advanced Studies in Architecture and Digital Fabrication (MAS DFAB),
Interactive Robotic Plastering (IRoP),
ETH Zurich,
2021

opposite: Interactive Robotic Plastering (IRoP) combines interactive design tools, an augmented-reality interface and robotic plaster spraying on site. An interactive computational model is used to translate the data from a motion-tracking system into robotic trajectories using design and editing tools and an audio-visual guidance system for in-situ projection. Project team: Daniela Mitterberger, Selen Ercan Jenny, Petrus Aejmelaeus-Lindström, Ena Lloret-Fritschi, Lauren Vasey, Eliott Sounigo, Ping-Hsun Tsai and David Jenny. Industry partners and sponsors: Eberhard Unternehmungen and Giovanni Russo.

Making the Past Productive

The issue also looks back in history with a compelling theme that unveils the extraordinary surge of technological advances in concrete construction. Although the 21st century presents us with a multitude of materials, constructions, building methods and processes, Silke Langenberg, Sarah Schlachetzki and Robin Rehm from the Construction Heritage and Preservation group at ETH Zurich discuss how the inventive spirit of the Industrial Revolution resulted in numerous innovations that have shaped our modern era and played a pivotal role in revolutionising contemporary construction practices. They highlight how digital fabrication presents opportunities now to rethink existing inventions developed in times of crisis and scarcity, which could also help to address today's pressing challenges in the face of climate change and resource shortages – fostering new opportunities to make the past productive by developing the undiscovered potential of historical patents in architecture.

Ultimately, this 𝘋 issue explores the evolving relationship between sustainability, technology and design in the construction industry. It provides a comprehensive narrative beyond mere theoretical discussions, presenting tangible projects that embody sustainability principles and utilise digital fabrication as a catalyst for change. Together, the articles aim to inspire a broader conversation about the future trajectory of architecture and construction, emphasising the role of technology in navigating the complexities of our built environment. With the topics that they bring into the spotlight, a new beginning is foreseen in understanding what realistic objectives can be achieved with holistic approaches. Demonstrating how digital fabrication is put to work towards achieving urgent sustainability goals within the context of building construction, it is hoped to inspire the next generation of researchers and practitioners to take the leap. 𝘋

Text © 2024 John Wiley & Sons Ltd. Images: p 6 © NCCR Digital Fabrication, photo by Roman Keller; pp 8–9 © Fabrication and Material Aware Architecture (FMAA), Università della Svizzera italiana (USI); pp 10, 12–13 © Gramazio Kohler Research, ETH Zurich; p 11(l) Centre for Building Technology and Process, Zurich University of Applied Sciences (ZHAW) and Fabrication and Material Aware Architecture (FMAA), Università della Svizzera italiana (USI), photo David Baumgartner; p 11(r) Centre for Building Technology and Process, Zurich University of Applied Sciences (ZHAW) and Fabrication and Material Aware Architecture (FMAA), Università della Svizzera italiana (USI), photo David Jenny

Mario Carpo

The Sustainable Lightness

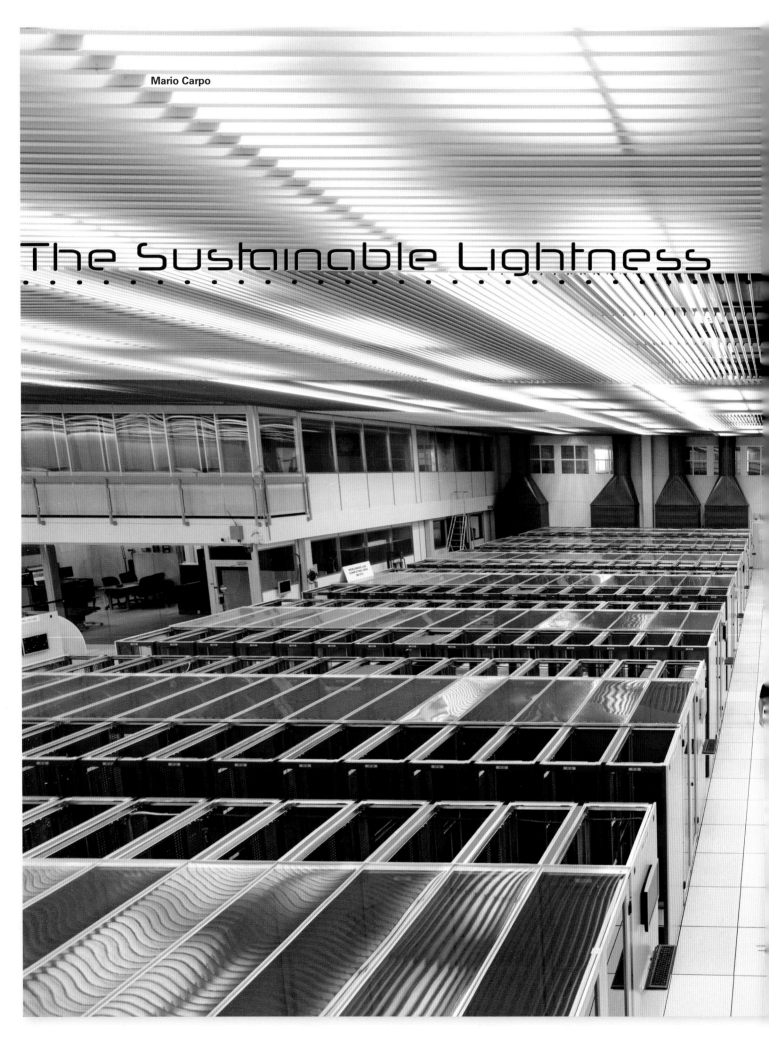

of Digital Fabrication

SPAN / Matias del Campo and Sandra Manninger, Urban Farming in Vienna 2123, 2023
above: With the rise of new forms of machinic intelligence, today's technological innovations may challenge our traditional understanding of the way we create texts, images and three-dimensional objects. Commissioned by Rhitz, the Vienna 2123 project attempts to leverage the predictive qualities of AI. Using off-the-shelf solutions such as ChatGPT and MidJourney, a future scenario for the City of Vienna is developed. The proposal considers aspects such as climate change, urban farming, climate migration and a changing urban ecology post-climate crisis.

The CERN (European Organization for Nuclear Research) Data Center, Geneva, in 2024
previous spread: An example of a large-scale data centre, at CERN, where 450,000 processor cores and 10,000 servers run 24/7.

Technical progress has bad press these days. No surprise in that, as today's anti-tech mood follows a recurrent pattern that has already run its course many times since the start of the Industrial Revolution: technical change tends to solve old problems by creating new ones; at times we focus on the old problems that are solved and at times on the new problems that arise. There are many excellent reasons to be wary of technical change today, and I shall not even need to review them, because they are often self-evident, and ubiquitous in the media, in academia, and in every educated conversation. Technical change, since the spring and summer of 2022, has meant the unexpected resurrection of artificial intelligence, under the guise of generative AI. Generative AI has stunned the general public not less than the professionals and scientists that should have seen it coming – and, with few exceptions, didn't. As a result, when opinion makers, philosophers, industrialists and politicians mention technology today, it is often not to praise its potential but to

It seems, in many quarters, that technology is viewed as bad for aspirations for a sustainable future. This Neo-Luddite narrative is hampering the way the advantages and disadvantages of digital technology are perceived. Mario Carpo, Reyner Banham Professor of Architectural History and Theory at the Bartlett School of Architecture, University College London (UCL), argues that new technologies succeed older ones because they perform tasks better and more efficiently, innovatively and faster, both changing and responding to changes in societal and urban realms.

warn of its risks – risks that authoritative and even official government sources have recently ranked on a par with other 'societal-scale risks' like pandemics and nuclear wars.[1]

Such alarming socio-technical prospects are for the most part a knee-jerk reaction following the sensational release of the first AI-powered chatbots. Understandably, a similar aura of suspicion and fear has now spread to include even the most pedestrian and utilitarian applications of artificial intelligence and applied robotics, like those that might be of use to the design professions. Compounding and amplifying similar neo-Luddite sentiments, however, another powerful prejudice also appears to be at play here – one which is particularly relevant to the building and construction industry. This is the claim, endorsed by many almost as an article of faith, according to which computational technologies should be seen as environmentally unfriendly – or even worse, as the most environmentally unsustainable technology of all.

Taking into account the full life-cycle assessment of today's tools for electronic computing, the sourcing of the raw materials, and the staggering amounts of energy needed for building and sustaining the machineries and the infrastructure that keeps computers running and connected to one another, it is indeed easy to show that digital technologies today are 'environmentally devastating'.[2] As I started to write this piece, a few days ago, I stumbled upon an article by the generally well-informed tech columnist of *The Observer*, the respected academic and scholar John Naughton, titled 'Why AI is a Disaster for the Climate', who cites a study proving that the carbon footprint produced by training a single large language model (LLM) is the equivalent of that left by 125 round-trip flights from New York to Peking.[3] That is good to know; but then one should also ask how many round-trip flights from New York to Peking will be avoided thanks to the services that this specific LLM will provide over time.

One technology replaces another when it is better than the technology it replaces – when it isn't, it doesn't. That's the Darwinian logic of technical change and that's not rocket science; it's double-entry bookkeeping. Ideology often obfuscates simpler calculations of profits and losses; in this case, however, the profit and losses are not counted in cash, but in environmental loads, so this calculus has somewhat wider implications – for us all and for humankind. This is why ideology and prejudice are even more dangerous in this instance than they normally tend to be. For, if we are not made blind by ideology, we should be able to see that digital tools do not extend but reverse the mode of functioning of the mechanical tools they replace. The technical logic of digital mass-customisation is the opposite of the technical logic of industrial mass-production. The technical logic of digital communications is the opposite of the technical logic of mechanical transportation. Both arguments have been known for a long time and they are, in theory, largely accepted and uncontroversial; their practical consequences, however, are momentous – in general, and even more so when we design and build architecture and cities.

Phasing Out Mass-production and Mass-transportation

Mass-production, the foundation of the Industrial Revolution, is based on mechanical matrixes (cast, dies, stamps or moulds) of which the upfront cost must be amortised by repeated use, hence the iron law of industrial modernity: the more identical copies of the same model we make, the cheaper each copy will be. But digital fabrication for the most part is not matrix based, hence identical reproduction in the digital world does not generate saving. In a digital design-to-fabrication process, making more identical copies of the same model will not make any of them any cheaper: if we are in the business of 3D-printing teapots, for example, the cost of each teapot will always be the same, no matter how many we make. Digital mass-customisation means that, in theory, we could make any number of identical or non-identical items at the same unit cost; when digitally made, each item is an individual one-off, even when serially reproduced.

In the old, industrial system, bigger used to mean cheaper: bigger factories meant lower costs; bigger markets meant cheaper goods. But in the new, post-industrial system, economies of scale do not apply, hence scale in theory does not matter – neither the

> Digital mass-customisation means that, in theory, we could make any number of identical or non-identical items at the same unit cost

scale of our production chain, nor the scale of the sales we need to break even, nor scale of our catchment markets. To go back to the same example: in the old system, the cheapest of all imaginable teapots would have been, ideally, a global standard, mass-produced in as many identical copies as possible in the biggest possible factory located wherever production costs happened to be the lowest. In the new system, that centrally made and globally traded teapot would not be cheaper than a custom-made teapot 3D-printed in a robotic micro-factory just across the street – made only when needed, as needed, where needed; made on site, to specs, on demand, using local materials, robotic tools and locally produced renewable electricity. And this without even considering the economic and environmental costs of transportation, which in the old systems were deemed irrelevant.[4]

For the last one hundred years, entrepreneurs, designers and engineers have tried – with uneven results – to industrialise our retrograde, artisanal ways of building. That meant, first and foremost, replacing natural building materials with factory-made ones. Unlike natural materials, by definition variable, random and unpredictable, factory-made materials can be made compliant with reliable industrial standards. This is what engineers needed to rationally calculate, cost, design and build new and daring, unprecedented structures. This is why natural stones, for example, had to be transformed into an artificial stone which we call concrete; and natural wood had to be transformed into engineered wood – plywood, particle boards, cross-laminated timber – which is derived from timber, but is made in factories. In retrospect, this may well have been the right thing to do one hundred years ago. But it is not now. For one thing, we now know that this mode of production has no future, because at some point raw materials will run out, and the environmental loads created by their industrial transformation and global transportation are increasingly seen as unsustainable. But also, crucially, today's digital technologies *do not need* to standardise and mass-produce building materials anymore, because digital making does not work that way.

Fabrication and Material Aware Architecture (FMAA) group,
Foldcast,
Academy of Architecture, Università della Svizzera Italiana (USI), Mendrisio, Switzerland, 2022–4
opposite: Prefabrication is often associated with uniform and repetitive architecture. In this context, it is challenging to incorporate more complex, custom shapes such as those that are typically necessary for optimised structures. Foldcast introduces a paper-based, flexible, nonstandard formwork without notable extra costs, allowing customised concrete elements to be realised through digital fabrication.

alterfact,
One of a Kind,
2015
above left: Machine production enters in dialogue with traditional craft in a series of vases based on an identical 3D model and reproduced with a clay 3D printing process. Sydney-based design studio alterfact specifically uses glitches in the printing process to explore making each object one of a kind – a beautiful example of how mass-customisation based on the tools of the Third Industrial Revolution extends the positions of the craftsperson in relation to handmade objects.

Zach Cohen,
Dripping,
2018
above right: The raw construction material meets computational design and intelligent robotic labour. In the dripping workflow, Zach Cohen explores digital fabrication to reimagine both the immaterial and physical labour of architectural design. The designer programs the robot to deposit concrete at specific points in space, or 'drops'. Each drop has two temporal parameters that are encoded within it: extrusion time and wait time, which determine local structure and aesthetics.

Robots Can Revive Ancestral Ways of Making

Using today's digital technologies, we can use any variable building material – as found, if need be – because each, even minute, building component can be individually listed and handled, and the assembly of any number of parts optimised in the pursuit of any given design task. When dealing with randomised components, or with the irregular assembly of regular parts – bricks, for example – today's computational building technologies can already achieve, in theory, a level of granular precision far exceeding that of any expert pre-industrial artisan, but faster and cheaper. Pre-industrial artisans often did not have a choice; living, as they did, in a world of physiocratic penury, they had to make do with whatever building material was locally available. But they also knew how to deal with each different log, plank or piece of timber they could put their hands on: literally, by handling it, they would know where in the building it should go, or how to use it at best. Today's robotic vision can already emulate some of those ancestral skills, even reviving some of the 'circularity' that was inevitable in material cultures where labour was proportionally more abundant and cheaper than raw materials. The Industrial Revolution reversed the terms of that cost equation, but we are now getting back to where we stood for centuries: raw materials are getting scarce again, and intelligent robotic labour will soon be available in almost unlimited amounts – and at zero costs, or at negligible marginal and environmental costs.

We may easily imagine that using machine learning (indeed, even generative AI), a robot could soon be able to scan an open pasture, pick and choose the best boulders for whatever task it was given, break and resize them if needed for geometrical congruence, and use all and only the resulting bits and pieces to build a close-fitting drystone wall.[5] And that robotic job wouldn't even need to be scripted: a dataset of examples of local drystone walls could be enough to train the model. A stone, if kept in the same place where nature has put it, has definitely no carbon footprint. A tree, if converted into a building within walking distance from where it grew, is a carbon negative – that is, it sequesters carbon dioxide, so long as it doesn't burn or rot.

Gramazio Kohler Research
in collaboration with the
Robotic Systems Lab, Chair
of Landscape Architecture
and Vision for Robotic Lab,
ETH Zurich,
Circularity Park,
Eberhard Unternehmungen AG,
Oberblatt,
2021-22

right: The robotically constructed retaining wall was built with an adaptive planning and fabrication pipeline that steers construction towards digitally defined global geometries while allowing for the use of abundantly available – and highly varied – locally sourced rock and recycled demolition materials that are extremely low in embodied energy. The pipeline relies on a parallelised planning algorithm and custom software interface that combines feature-based candidate seeding with heuristics adapted from traditional masonry methods, constrained registration, rigid body simulation and learned classifiers. Project team: Ryan Luke Johns and Lauren Vasey.

ZHAW Centre of Building Technology and Process and USI
Fabrication and Material Aware Architecture (FMAA) group in
collaboration with Oxara,
'earth to earth' Winter School,
Università della Svizzera Italiana (USI),
Mendrisio, Switzerland,
2023

opposite left: Earth goes back to earth: a robotic 3D-printing process was used to produce thin, clay formwork elements to build an outdoor staircase, casting a special earth-based concrete. As a conceptual counter-position to current building practice, the proposed structure is built taking its predicted decay into account. Project team: David Jenny, Ena Lloret-Fritschi, Yen-Fen Chan, Sacha Cutajar, Fabio Amicarelli, Selen Ercan Jenny, Samuel Rebelo Garcia and Patric Fischli-Boson.

We all know we cannot build office towers and high-rises that way, so some trade-off with carbon-heavier ways of building will be necessary. On the other hand, chances are that we may not need to build so many new office towers in the near future. After all, at the time of writing (January 2024) many office buildings have been empty for most of the last four years. During the Covid-19 pandemic we didn't use office buildings at all, yet most office work kept being done. And if we want to talk about waste, what about building entire parts of our cities that are designed to be used only a few hours a day on some days of the week, and are kept empty the rest of the time?

The long and short is, we have now conclusive evidence to suggest that digital mass-customisation can be cheaper, faster, smarter and more environmentally sustainable than the mechanical mass-production of standardised industrial goods; just like the electronic transmission of information is cheaper, faster, smarter and more environmentally sustainable than the mechanical transportation of people and things. The technical logic of the Industrial Revolution was based on the exploitation of human labour and of natural resources. We always knew that the exploitation of human labour was unjust. We now also know that the unlimited exploitation of limited natural resources will soon have to end. Intelligent robotic labour is an alternative to, and can be a way out of, the technical logic of industrial modernity. ⌖

Notes
1. See Center for AI Safety, 'Statement on AI Risk: AI Experts and Public Figures Express Their Concern About AI Risk', 30 May 2023: www.safe.ai/statement-on-ai-risk, and the international conference summoned by the UK government at Bletchley Park on 1–2 November 2023: www.gov.uk/government/publications/frontier-ai-capabilities-and-risks-discussion-paper.
2. Christina Cogdell, *Toward a Living Architecture? Complexism and Biology in Generative Design*, University of Minnesota Press (Minneapolis, MN), 2018, pp 95–101.
3. John Naughton, 'Why AI is a Disaster for the Climate', *The Observer*, 23 December 2023: www.theguardian.com/commentisfree/2023/dec/23/ai-chat-gpt-environmental-impact-energy-carbon-intensive-technology#:~:text=Basically%2C%20because%20AI%20requires%20staggering,electricity%20at%20a%20colossal%20rate.
4. Mario Carpo, *Beyond Digital: Design and Automation at the End of Modernity*, MIT Press (Cambridge, MA), 2023, pp 4–33.
5. See recent research carried out at ETH Zurich by the team of Fabio Gramazio and Matthias Kohler, in particular Ryan Luke Johns's doctoral thesis 'Autonomous Dry Stone: Mobile Robotic Construction with Naturally Nonstandard Materials', 2023: www.research-collection.ethz.ch/handle/20.500.11850/660060.

Gramazio Kohler Research in collaboration with the Master of Advanced Studies in Architecture and Digital Fabrication, Interactive Robotic Plastering (IRoP), ETH Zurich, 2021

above right: IRoP introduces a system enabling designers and skilled workers to engage intuitively with an in-situ thin-layer plaster printing technique (robotic plaster spraying). A customisable computational toolset converts human intentions into robotic motions while respecting robotic and material constraints. Project team: Daniela Mitterberger, Selen Ercan Jenny, Petrus Aejmelaeus-Lindström, Ena Lloret-Fritschi, Lauren Vasey, Eliott Sounigo, Ping-Hsun Tsai and David Jenny.

Text © 2024 John Wiley & Sons Ltd. Images: pp 14–15 © CERN; pp 16–17 © SPAN / Matias del Campo and Sandra Manninger. AI-generated image created with ChatGPT and Midjourney; p 18 © Fabrication and Material Aware Architecture (FMAA), Università della Svizzera italiana (USI); p 19(l) © alterfact; p 19(r) © Zach Cohen; p 20, 21(r) © Gramazio Kohler Research, ETH Zurich; p 21(l) Centre of Building Technology and Process, Zurich University of Applied Sciences (ZHAW) and Fabrication and Material Aware Architecture (FMAA), Università della Svizzera italiana (USI)

Timothy Wangler,
Yamini Patankar and
Robert J Flatt

Sustainable Digital Concrete
Myth, Reality or Emerging Opportunity?

National Centre of
Competence in Research
(NCCR) Digital Fabrication,
DFAB House concrete
innovation objects,
NEST, Empa Campus,
Dübendorf, Switzerland,
2018

Completed in 2019, the DFAB House is a demonstration module from the NCCR Digital Fabrication, a Swiss research consortium. Three concrete innovation objects can be seen here: the Mesh Mould Wall by Gramazio Kohler Research, a double-curved structural wall produced robotically; the Smart Slab by Digital Building Technologies, a structurally optimised prestressed slab with integrated systems produced by 3D-printing the concrete formwork; and bespoke façade mullions produced by Gramazio Kohler Research and the group of Robert J Flatt, ETH Zurich, the Smart Dynamic Casting (SDC) process.

Timothy Wangler, Robert J Flatt and Yamini Patankar,
Concrete Composition and Environmental Impact,
ETH Zurich,
2024

Concrete's reputation in construction as unsustainable stems from its enormous production. However, in comparison with many other materials, when normalised by the relative production amounts, it is actually relatively eco-friendly. Thus incremental improvements, such as through material efficiency, can have potentially large impacts. Illustrated on the left, a typical cement plant; and on the right, the roof of the aircraft hangar in Orvieto, Italy (1935), designed by Pier Luigi Nervi.

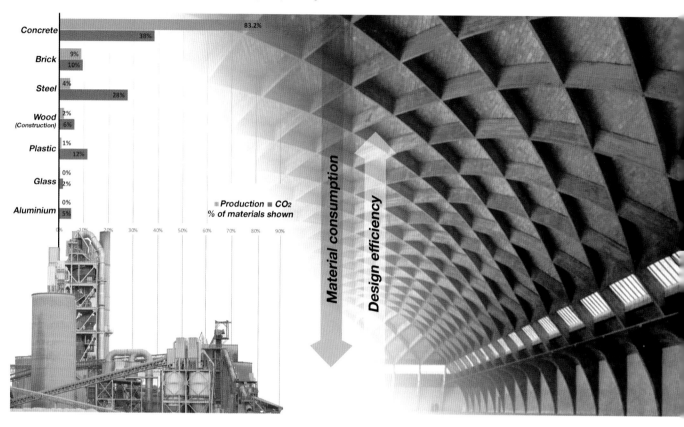

Concrete forms the backbone of modern infrastructure and architecture, playing this incredible role in both invisible and highly visible ways. Its omnipresence is staggering when one pauses to consider it – around 14 cubic kilometres (3.4 cubic miles) of concrete are produced per year,[1] which means a 1.75-metre (5 foot 9 inch) cube of concrete per person is produced annually. Considering current trends, the total mass of concrete ever produced by humanity will exceed all biomass on the Earth's surface by 2040.[2] Welcome to the Anthropocene.

Why this ubiquity? There are very good reasons for it, and they have to do with the raw materials and the properties of the product. Concrete requires three essential ingredients, which are locally available and easily produced worldwide: water, cement and aggregate (sand and gravel). Properly proportioned, they create a malleable fluid that later hardens and durably bears a structural load. Mixing water and cement initiates a complex series of chemical reactions leading to this liquid-to-solid transformation, making a paste that glues the aggregates together upon hardening. This is essentially the most efficient means ever devised of producing artificial stone, and similar to stone, concrete is brittle and weak in tension. Its formability is yet another superpower here, enabling it to be combined with steel and harness the strengths of both materials. The union of these two materials is really what forms the backbone of today's infrastructure. Steel provides tensile strength and ductility; concrete provides compressive strength and corrosion protection.

And yet, in recent years concrete has carried a negative reputation. An award-winning article in the *Guardian* newspaper notably proclaimed it to be 'the most destructive material on earth'.[3] The staggering scale of concrete production would seem to make this true by default, resulting as it does in concrete's responsibility for anywhere between 6 and 8 per cent of global CO_2 emissions. However, the problem is not concrete; it is cement. More precisely, it is clinker: the major reactive, man-made part of modern cement, produced by burning limestone and clay to bring about new mineral phases at high temperatures (approximately 1450°C (2640°F)). CO_2 is present in limestone and is released in the first stage of this process (about 60 per cent of concrete's carbon emissions), while the high temperatures require fuel that releases even more CO_2 (about 40 per cent).

Unbelievably huge amounts of concrete, a major contributor to our contemporary world, are produced every year, and this continues to increase exponentially. ETH Zurich researchers **Timothy Wangler, Yamini Patankar and Robert J Flatt** explain the pros and cons of the digital fabrication of concrete, and the research still to be done in this relatively young and experimental subset of the construction industry and material science.

Although clinker may be the 'poison pill' in concrete, it is also its most expensive element, so economic incentive has effectively diluted its presence via efficient packing of cheap, low-carbon-impact aggregates, as well as partial substitution with other reactive, lower-carbon-impact 'supplementary cementitious materials'. Considering this in proportion to the amount of concrete produced, concrete compares extremely favourably – in fact by an order of magnitude – to other materials such as brick, steel, aluminium and even timber.[4]

Considering all this, one comes to the conclusion that concrete is in fact a sustainable, irreplaceable material, but with room for improvement.[5] Given its scale, even minor changes in usage produce major impacts. Reducing clinker means either decreasing its content in a given volume of concrete or reducing the total amount of concrete, and it is the latter where digital fabrication has potential. Let us therefore examine the recent revival of automation in construction and digital fabrication with concrete, beginning with its primary motivation: construction productivity.

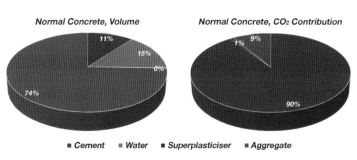

Yamini Patankar, Robert J Flatt and Timothy Wangler,
Concrete Comparative CO_2 Impact versus Usage,
ETH Zurich,
2024
Concrete is produced by mixing water, cement, aggregates (sand and gravel) and other additives. While forming a small portion of the concrete, the cement dominates the overall carbon impact.

Productivity

It is well known that architecture, engineering and construction (AEC) have lagged in productivity for decades when compared to other sectors such as manufacturing. Productivity, defined as cost of construction per unit of time, has even declined, with project cost overruns and delays being the norm. Numerous complex reasons contribute to this, but a significant one is that the automation revolution which benefited the manufacturing sector has apparently left AEC behind, giving the more recent digital revolution a unique opportunity to reshape things for AEC.[6] Industrial robots dominate factory floors, while construction sites still resemble those of a century ago, crawling with construction workers performing manual, back-breaking jobs that are among the most unsafe in many countries. An ageing workforce performs these jobs, younger workers refuse to take them up, and immigration policies struggle to fill the gap, leading to an overall labour shortage and increased labour costs.[7] This has cascaded into exacerbation of numerous societal issues, especially the housing crisis across numerous countries,[8] as skilled labour migrates towards higher-paying commercial or infrastructure construction and away from residential. The housing crisis was even identified as a primary reason for recent civil unrest in Ireland.[9]

An oft-named solution to these crises is an increase in automation, digitisation and a more modular, factory-based approach to construction.[10] The cost of a given construction consists of two major parts: labour and materials. A 2007 *STRUCTURE* magazine study indicates that in concrete construction, formwork labour forms the highest portion of cost, and this continues to increase.[11] Alternative methods, such as using concrete masonry units (CMUs) or brick, remain labour-intensive and display a similar cost breakdown. Digital fabrication technologies offer an obvious solution to reduce these costs – a study published in 2019 showed explicitly that 3D-printing concrete was profitably replacing either formwork or masonry labour.[12] Additionally, digital fabrication technologies can potentially be more easily streamlined, reducing delays. While these incentives are really at the heart of the revival of digital fabrication with concrete, there is also a golden opportunity for more sustainable construction.

Yamini Patankar, Robert J Flatt and Timothy Wangler,
Material impact of structures,
ETH Zurich,
2024

above: While conventional construction uses concrete with a lower impact than construction with digital fabrication, digital fabrication allows more economical production of material-saving, structurally efficient designs reminiscent of Pier Luigi Nervi's. While requiring higher cement content in the material, the overall material usage can be decreased. Conventional construction is represented here (on the left) by the Chandigarh Architecture Museum in India (1997), designed by Shivdatt Sharma; and digital fabrication (right) by Gramazio Kohler Research's Eggshell Pavilion at the Vitra Design Museum, Weil am Rhein, Germany (2022).

Yamini Patankar, Robert J Flatt and Timothy Wangler,
Costs of construction complexity,
ETH Zurich,
2024

opposite top: Conventional construction has a high cost due to high labour costs and high labour requirements for formwork. Digital fabrication reduces the proportion of labour cost significantly, and does so for both simple and complex elements. (Pie chart information from García de Soto *et al.*, 'Productivity of Digital Fabrication in Construction': https://doi.org/10.1016/j.autcon.2018.04.004.)

Sustainability

Reducing clinker's environmental impact can be done by reducing the concrete footprint (CO_2 per unit of mass), reducing the total amount of concrete in construction, or both. Extending a building or structure's service life might also be viewed as reducing environmental impact. However, when one seriously examines the material footprint of concretes used in digital fabrication, one finds such concretes often have double the cement contents than ordinary ones.[13] The durability has also been called into question, although research is still in its early stages. Given this, how can digital fabrication with concrete ever be considered sustainable?

Digital fabrication can reduce construction's carbon impact only through material-efficient design. For exactly the same reasons that it is being adopted with respect to productivity – rising labour costs – digital fabrication can also offer a path back to complex, shape-efficient construction, which has essentially been priced out of existence. For example, Pier Luigi Nervi's ribbed slab designs, exemplified in the Gatti Wool Factory in Rome (1953), are extraordinarily costly to produce with today's conventional techniques. Considering that slabs constitute anywhere from 40 to 60 per cent of the concrete in typical three- to eight-storey buildings, this represents an enormous potential for reduction of carbon footprints.[14] A signature project of the Swiss National Centre of Competence in Research (NCCR) Digital Fabrication is the DFAB House (2019), a residential unit at the NEST modular research and innovation building in Dübendorf, Switzerland. It has been constructed using digital fabrication techniques, showcasing a topologically optimised 'Smart Slab' from Digital Building Technologies, as well as a double-curved wall from the Mesh Mould project and materially optimised façade mullions from Smart Dynamic Casting, both by Gramazio Kohler Research. Another instance is the HiLo roof (2022), a double-curved thin concrete shell over a residential unit also at the NEST building, digitally designed and produced by the Block Research Group.

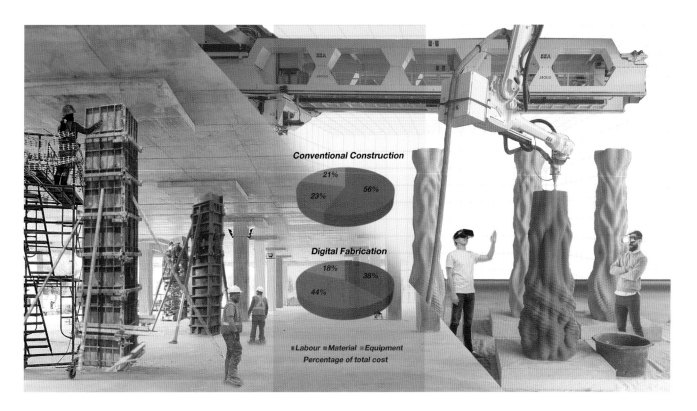

Robert J Flatt,
Relation showing environmental impact of concrete construction,
ETH Zurich,
2024

right: The environmental impact of a concrete construction per year of service life can be qualitatively estimated by multiplying its carbon footprint by the total amount of concrete used and dividing by the service life. Reducing environmental impact can be envisioned as reducing concrete footprint, reducing total mass, or both.

$$\text{Environmental Impact of Concrete per Year of Service} \propto \frac{\frac{\text{Embedded } CO_2}{\text{Unit Mass}} \times \text{Total Mass}}{\text{Service Life}}$$

Block Research Group,
HiLo concrete roof,
Dübendorf, Switzerland,
2022
above: The HiLo concrete roof is a double-curved sandwich structure which features two thin layers of reinforced concrete with insulation blocks in between. It utilises its geometry in combination with the lightweight structure to create a large unobstructed space of 120 square metres (1,300 square feet). Rather than use expensive single-use formwork, it was created using a flexible formwork made from a cable-net and membrane system.

Timothy Wangler, Robert J Flatt and Yamini Patankar,
Cement content, productivity and sustainability,
ETH Zurich,
2024
right: Concrete in its liquid form consists of aggregates (sand and gravel) and fluid cement paste (water and cement). As cement paste increases, ease of processing increases, but so does cement content, creating an apparent trade-off between productivity and sustainability.

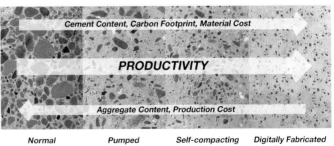

The Path Forward
So, can digital fabrication truly be considered sustainable with respect to carbon footprints? At this point in time, no. Most industrial realisations of 3D concrete printing, the most visible digital fabrication technology for concrete, have essentially been the replacement of concrete masonry unit walls with a 3D-printed analogue. In such cases, the printed concrete rarely functions structurally and typically acts as a lost formwork, where steel reinforcement is inserted into printed voids and a structural concrete is later cast. Unlike concrete in conventional construction, which is often optimised for the lowest cement contents possible, printed concrete is high in cement content, making it more expensive and with a higher carbon footprint. This higher cement content stems from higher paste contents needed to ease processing – high aggregate contents make processing difficult. Additionally, digital fabrication requires rapid early strength. Because this typically increases with cement content, most printed concrete therefore contains more cement than required for the final application, resulting in overdesign of most printed structures. In spite of these drawbacks, industry interest in digital fabrication still remains high because the trade-off of increased cement content is perceived to provide enough productivity benefit.

This gives the impression that adopting digital fabrication with concrete is a choice to increase productivity at the expense of sustainability. A closer look, however, suggests that this is a trend that has been progressing in concrete construction, somewhat unnoticed, over time. From the material footprint standpoint, higher paste contents, which correlate to higher cement contents, are required for the major concrete productivity innovation of pumping. Even higher are the paste contents in self-compacting concrete, a more recent innovation that speeds up construction by eliminating the need for vibration. Digital fabrication technologies such as 3D printing are apparently the latest iteration in this trend. However, this trade-off between productivity and sustainability may miss more substantial opportunities.

Indeed, digital design and fabrication, which has been viewed as a replacement for human labour until now, can also be regarded as an enabling technology for more creative, materially efficient designs – a door opener to a material-efficient architecture no longer constrained by the cost imperatives of traditional labour-intensive construction practice. While this may come at the expense of a higher material footprint, the potential sustainability benefit of a reduced total amount of concrete can make the overall structure more sustainable, which is where new opportunities for design are seen. The potential sustainability benefits do not stop there – a structurally efficient concrete construction reduces not only concrete but also reinforcement steel, another contributor to CO_2 emissions. Functionally graded structures can be more easily produced, bio-inspired architectural designs more easily realised across scales, and innovative systems more easily integrated to reduce energy consumption over a building's lifetime. The current revolution in the enabling technologies of AI and extended reality can be more easily incorporated with digital fabrication to push sustainable design even further, embracing notions of circularity through designs for both assembly and disassembly. The possibilities are endless, so digital fabrication should be viewed as a technology that expands and aids the design process for human creativity to have a greater lever on sustainability in AEC. Ultimately, for all of humanity's foibles, it is still our creative impulse that has served us best in getting out of difficult situations. Any technology that enables this creative impulse can and should be viewed as a potential asset – provided it is used judiciously and not cynically. ∆

Notes
1. Global Cement and Concrete Association, 'Cement and Concrete around the World': https://gccassociation.org/concretefuture/cement-concrete-around-the-world/, consulted 4 February 2024.
2. Brian Potter, 'There Will Soon Be More Concrete Than Biomass on Earth', Heatmap News, 8 March 2023: https://heatmap.news/economy/the-planet-s-jaw-dropping-astonishing-downright-shocking-amount-of-concrete.
3. Jonathan Watts, 'Concrete: The Most Destructive Material on Earth', *The Guardian*, 25 February 2019: www.theguardian.com/cities/2019/feb/25/concrete-the-most-destructive-material-on-earth; GNM Press Office, 'Guardian Wins at the Foreign Press Association Awards', 26 November 2019: https://www.theguardian.com/gnm-press-office/2019/nov/26/guardian-wins-at-the-foreign-press-association-awards.
4. Michael F Ashby, *Materials and Sustainable Development*, Butterworth-Heinemann (Oxford), 2022, pp 485–534; Geoffrey Hammond and Craig Jones, *Embodied Carbon: The Inventory of Carbon and Energy*, Building Services Research and Information Association (Bracknell), 2011, p 136.
5. Robert J Flatt, Nicolas Roussel and Christopher R Cheeseman, 'Concrete: An Eco Material That Needs To Be Improved', *Journal of the European Ceramic Society* 32 (11), August 2012, pp 2, 787–98.
6. Filipe Barbosa et al, *Reinventing Construction Through a Productivity Revolution*, McKinsey Global Institute (New York), 27 February 2017: www.mckinsey.com/industries/capital-projects-and-infrastructure/our-insights/reinventing-construction-through-a-productivity-revolution.
7. Turner & Townsend, 'International Construction Market Survey 2023': www.turnerandtownsend.com/en/perspectives/international-construction-market-survey-2023/.
8. US Government Accountability Office, 'The Affordable Housing Crisis Grows While Efforts to Increase Supply Fall Short', 12 October 2023: www.gao.gov/blog/affordable-housing-crisis-grows-while-efforts-increase-supply-fall-short; Ramzi Chamat, 'The Housing Dilemma in Switzerland and the Chronic Shortage', Oaks Lane, 13 November 2023: https://oakslane.ch/en/swiss-real-estate-news/the-housing-dilemma-in-switzerland-and-the-chronic-shortage-ramzi-chamat-oaks-lane-sa-array-2023-11-13; 'Germany's Government Calls Summit to Combat Housing Crisis', euronews, 25 September 2023: www.euronews.com/2023/09/25/germanys-government-calls-summit-to-combat-housing-crisis.
9. 'Dublin Riots Expose Irish Frustration at Housing, Cost-of-Living Crisis', *TIME*, 6 December 2023: https://time.com/6343248/dublin-riots-irish-youth-housing-crisis/.
10. Barbosa et al, *op cit*.
11. Robert H Lab, Jr, 'Think Formwork – Reduce Costs', *STRUCTURE*, April 2007: www.structuremag.org/?p=6141.
12. Eric L Kreiger, Megan A Kreiger and Michael P Case, 'Development of the Construction Processes for Reinforced Additively Constructed Concrete', *Additive Manufacturing* 28, August 2019, pp 39–49.
13. Robert J Flatt and Timothy Wangler, 'On Sustainability and Digital Fabrication with Concrete', *Cement and Concrete Research* 158, August 2022, 106837.
14. Jaime Mata-Falcón et al, 'Digitally Fabricated Ribbed Concrete Floor Slabs: A Sustainable Solution for Construction', *RILEM Technical Letters* 7, 2022, pp 68–78.

Text © 2024 John Wiley & Sons Ltd. Images: pp 22–3 © NCCR DFAB, photo Roman Keller; p 24: left © carterdayne/Getty Images, right © MAXXI Fondazione, MAXXI Museo nazionale delle arti del XXI secolo, Roma. Collezione MAXXI Architettura. Archivio Pier Luigi Nervi F4268. Composite image by Yamini Patankar; pp 25, 28(b) Images by Yamini Patankar, Timothy Wangler and Robert J Flatt; p 26: left @ Steve Speller / Alamy, right © Gramazio Kohler Research, ETH Zurich, photo Yen-Fen Chen. Composite image by Yamini Patankar; p 27(t): background construction site © Channarong Jaisan / Shutterstock, man in VR glasses (left) © Ground Picture / Shutterstock, man in VR glasses (right) © PanicAttack / Shutterstock, right © Axel Crettenand, Digital Building Technologies, ETH Zurich. Composite image by Yamini Patankar; p 27(b) © Robert J Flatt; p 28(t) © Block Research Group, ETH Zurich, photo Roman Keller

INJECTION 3D

Norman Hack, Harald Kloft, Inka Mai, Pierluigi D'Acunto, Yinan Xiao and Dirk Lowke

FROM STRUCTURAL

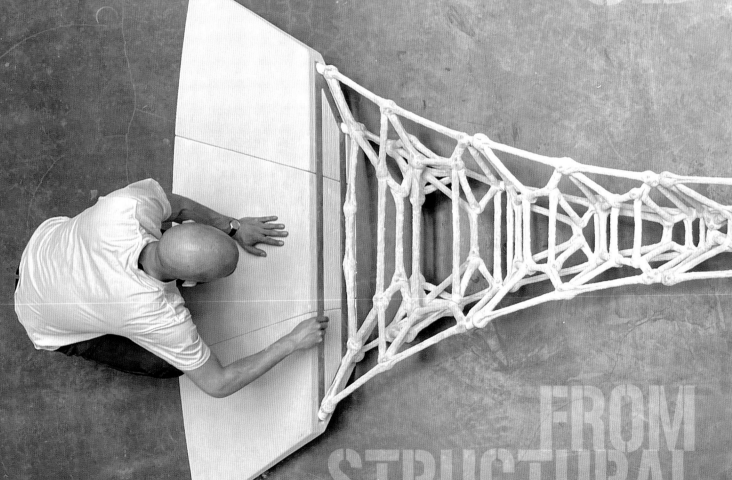

CONCRETE PRINTING

Yinan Xiao,
Injection 3D-printed bridge prototype,
Institute of Structural Design, Faculty
of Architecture, Civil Engineering and
Environmental Sciences,
TU Braunschweig, Germany,
2023
The assembled bridge spans 3 metres (9.8 feet) and consists of five individual components. Only compression forces act in the bridge, which are transferred to the solid 3D-printed bases.

GEOMETRY TO FABRICATION

Architects and engineers have historically developed and reinvented concrete's technologies, formwork and aesthetics to suit the pragmatic and philosophical aims of their times. Architect and computational design researcher **Norman Hack** and his co-authors introduce a contemporary method for the fabrication of concrete structural elements using the Injection 3D Concrete Printing (I3DCP) process formulated at TU Braunschweig and discuss its benefits over other types of printed concrete.

In his essay 'The Material Without a History' (2006), Adrian Forty, Professor Emeritus of Architectural History at the Bartlett School of Architecture, University College London (UCL), embarks on a compelling exploration of concrete's discontinuous development over the past century. He puts forward the thesis that successive generations of architects and engineers have approached concrete construction as if it had no previous history, constantly reinventing its constructive logic and formal expression,[1] beginning with Auguste Perret's pioneering trabeated structures characterised by their monolithic columns, beams and frames, through Le Corbusier's innovative cantilevered slabs of the Dom-Ino House (1914–15), to the elegant, material-saving reinforced concrete shells of the 1940s, then the exposed raw concrete surfaces of brutalist architecture in the 1950s, and the smooth, polished exposed concrete buildings that became a trademark of Swiss architecture in the early 2000s.

More recently, a noteworthy leap in the trajectory of concrete's narrative has been marked by the introduction of 3D-printing technology, an innovation that imparts a novel tectonic logic and surface aesthetics to this multifaceted material. The essence of 3D printing with concrete lies in the layering of extruded concrete struts, stacking to fashion intricate three-dimensional components. However, this technological innovation comes with a unique set of challenges, primarily concerning the interaction between material and process.

In conventional concrete casting, the formwork plays the central role of shaping and supporting until the material has sufficiently hardened. In contrast, 3D printing necessitates a paradigm shift whereby the material itself has to assume control of these tasks. Failure to do so could result in the collapse of the 3D-printed structure under its own weight. This heightened demand on the material imposes certain constraints on shaping possibilities, typically resulting in the adoption of single- or moderately double-curved geometries with limited overhangs.

The Injection 3D Concrete Printing (I3DCP) process, jointly developed by the Institute of Structural Design (ITE) and the Institute of Building Materials, Concrete Construction and Fire Safety (iBMB) of the Technische Universität Braunschweig (TU Braunschweig), aims to overcome these limitations. Investigated at both process and material level,[2] it differs from conventional 3D-printing techniques by using a robotic system that injects concrete into a vessel filled with a non-curing, non-Newtonian carrier liquid.[3] Additionally, the liquid's advanced formulation allows for its repeated use across numerous printing cycles, enhancing efficiency and sustainability. The technique unveils entirely novel tectonic and formal potentials: 3D printing with concrete is no longer bound to layering for the creation of solid components; it can now take place along unrestricted spatial trajectories. This allows for the printing of intricate spatial structures, departing and exceeding conventional perceptions of concrete constructions.

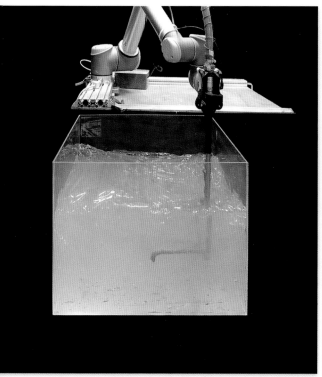 

Design Freedoms

The prospect of being able to print highly complex spatial patterns creates new design freedoms. In sustainable design, however, this freedom raises the critical question of which forms are meaningful and which are not. Examining this from a structural point of view, it is evident that concrete performs remarkably well in compression. In contrast, in tension it has only a fraction of its structural capacity, typically around one-tenth.

Graphic statics is a valuable basis for understanding these structural considerations. It was initially formalised in the second half of the 19th century when German engineer Carl Culmann developed a graphical method at ETH Zurich for the design and analysis of structures.[4] The method relies on two interdependent diagrams, namely the form diagram representing the structure's geometry with the external forces, and the force diagram showing the equilibrium of the nodes of the structure using closed cycles of force vectors. Unlike the conventional design-to-engineering approach, where a structure is first designed then calculated and iteratively optimised, graphic statics enables a direct design and interaction with the forces. One example of this approach is the Salginatobel Bridge in Switzerland, a masterpiece designed by Swiss engineer Robert Maillart in 1929.[5]

Yinan Xiao,
The Injection 3D Concrete Printing (I3DCP) process,
Institute of Structural Design,
Faculty of Architecture, Civil Engineering and Environmental Sciences,
TU Braunschweig, Germany,
2020

above left: A robot prints concrete into a non-hardening carrier liquid. The concrete remains stable in the suspension until it has hardened. A transparent sonographic gel was used to provide visual control. A reusable, cost-effective and robust suspension based on limestone powder was later developed.

Robert Maillart,
Salginatobel Bridge,
Schiers, Switzerland,
completed 1929

above right: The Salginatobel Bridge is a three-hinged, 90-metre (295-foot) span reinforced-concrete-arch bridge with a hollow box cross-section. The form of the arch was designed by Maillart using graphic statics, taking into account both the self-weight of the bridge and the effect of moving loads on the deck.

> The interdependence between the form and force diagrams makes it possible to create complex geometric structures that transfer their loads under compression in a structurally efficient manner

One downside of Culmann's method in the past was the labour-intensive process of graphically drafting the vector constructions. However, what once demanded considerable effort on paper in 2D is now calculated by computers in three-dimensional space within milliseconds. This advancement, and more particularly the extension of traditional graphic statics into vector-based graphic statics (VGS),[6] is an ideal prerequisite for the I3DCP process, as 3D force vectors can seamlessly be translated into robotic print paths. The interdependence between the form and force diagrams makes it possible to create complex geometric structures that transfer their loads under compression in a structurally efficient manner. Yet there are two key challenges when designing such structures for the I3DCP: firstly, in creating print paths that avoid undesired intersection with the already printed concrete struts; and secondly, consideration must be given to the constraints of print space size, necessitating the segmentation or modularisation of components to facilitate the creation of large structures from small components.

Material Variables

The interplay between process and material is paramount in additive manufacturing. The I3DCP method introduces an additional layer to this interconnection, involving the intricate dynamics between the two distinct materials – concrete and the carrier liquid. Alongside the usual fabrication parameters such as nozzle diameter and extrusion rates, material variables including yield stress, plastic viscosity and density come into play.

A basic prerequisite for successfully implementing the I3DCP process is the accurate calibration of both materials' yield stress. If the carrier liquid's yield stress is excessively high, leading to rigidity, there is a risk of incomplete closure after the nozzle's passage, compromising the attainment of uniform cross-sections. Conversely, a fluid with low yield stress may struggle to uphold the position of the injected concrete, especially as the concrete's density is higher than that of the liquid, potentially resulting in sinking or deformation of the printed strand. Process-related errors, including over-extrusion causing material curling or under-extrusion leading to strand detachment, are also potential challenges. In the initial experiments at TU Braunschweig, a transparent sonographic gel was used as a carrier liquid, enabling visual inspection of the printing process. However, the concrete's high alkali content degraded the gel after a few prints. A non-transparent but more robust suspension based on limestone powder was therefore developed for subsequent experiments, costing only a fraction of the sonographic gel.

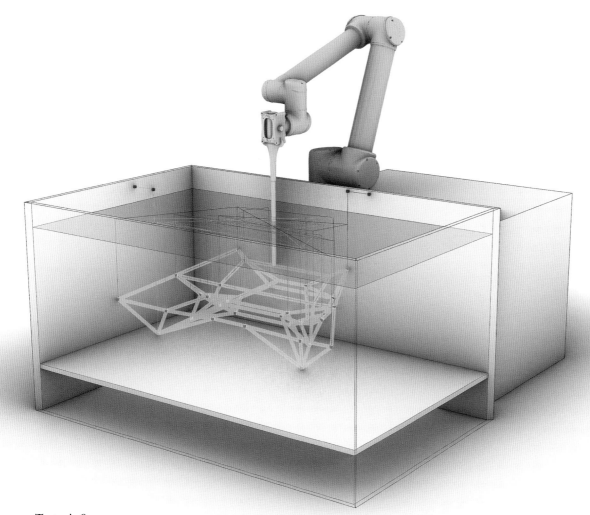

Yinan Xiao,
Print path simulation,
Institute of Structural Design, Faculty
of Architecture, Civil Engineering and
Environmental Sciences,
TU Braunschweig, Germany,
2023
Robotic print path for the fabrication of a bridge component using the I3DCP process. To save carrier liquid, components are modularised and printed in containers up to 0.75 cubic metres in size, taking into account that the nozzle does not cross any already printed elements during printing.

Tectonic Scope

Besides numerous smaller material samples and test specimens utilised for process calibration, I3DCP has been tested in two case studies. Both examples are compression-only structures that are not printed monolithically in a single large vessel, but as components that are later assembled into a coherent whole, showing how structural reinforcement and complicated connections between the individual components can be avoided.

The initial case study, a coffee table approximately 100 x 100 x 40 centimetres (40 x 50 x 15 inches) in size, was conceived in collaboration with students from TU Braunschweig in a design studio utilising VGS.[7] The table's geometry was derived through force diagrams based on a predefined structural topology, external forces and a support plane. The interplay between the form and force diagrams ensured an even distribution of inner forces among strut elements for optimal stress distribution. The table comprises three point-symmetrical components, each of which was printed within a 66 x 50 x 30 centimetre (26 x 20 x 12 inch) container filled with carrier liquid made of limestone powder. A 0.2-second pause at the strut joints, without halting material flow, created interfaces for connection in the subsequent assembly. Each strut was printed in approximately 180 seconds. After an 18-hour curing period, the components were lifted from the carrier liquid. By simply leaning the components against each other, the structure is based solely on compressive forces, which are transferred to the floor via 3D-printed supports.

Yinan Xiao and Noor Khader,
Injection 3D concrete printed table,
Institute of Structural Design, Faculty
of Architecture, Civil Engineering and
Environmental Sciences,
TU Braunschweig, Germany, 2021

The small coffee table was realised in a Digital Building Fabrication design studio. The printing time for each component was about three minutes. The three components lean together and form a stable compression-only structure.

Building on this smaller prototype, a second structure, a pedestrian bridge spanning a greater distance and comprising multiple, structurally more intricate components was designed.[8] Mirroring the table's design methodology, the global design of this larger structure was also conceived using VGS, but now integrating advanced fabrication constraints. The test results of the printed samples with different settings showed that certain parameter ranges warranted good printing results. These domains were then used as reference points for improving the design of the structure. The constraints included, for example, the number of struts merging at a joint, strut length, angles between the struts, distance between the struts, as well as the critical forces to prevent buckling.

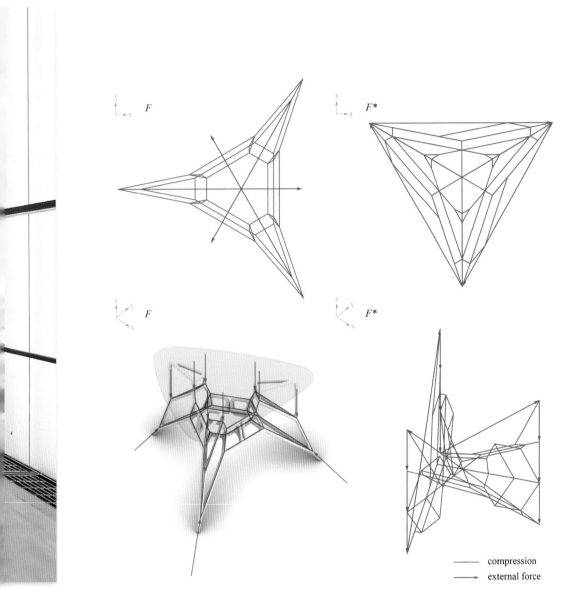

— compression
→ external force

Comprising five components, and designed to be built at a 1:5 scale, the bridge spans a length of 3 metres (9.8 feet) in a compression arch configuration, with forces seamlessly transmitted to the 3D-printed bases. The structural optimisation in terms of the constraints was individually implemented on each component to meet both I3DCP requirements and static equilibrium. Given the elevated precision requirements at the interfaces of the components, a high-precision solution for the connectors became imperative. To address this, a systematic workflow was developed: first, each of the five components was printed individually in containers up to 160 x 100 x 70 centimetres (63 x 40 x 27 inches) in size and filled with the reusable limestone powder carrier liquid. After the curing and extraction process, detailed 3D scans of the individual components were generated using a handheld infrared-based scanner.

Yinan Xiao,
Form and force diagrams of a coffee table,
Institute of Structural Design, Faculty of Architecture, Civil Engineering and Environmental Sciences,
TU Braunschweig, Germany,
2021
The table consists of three 3D-printed components made of concrete struts connected at nodes. The figure shows the form diagram (F) and force diagram (F*) of the table in the top view (upper row) and perspective view (lower row). The forces in the concrete struts are optimised to be evenly distributed throughout the structure.

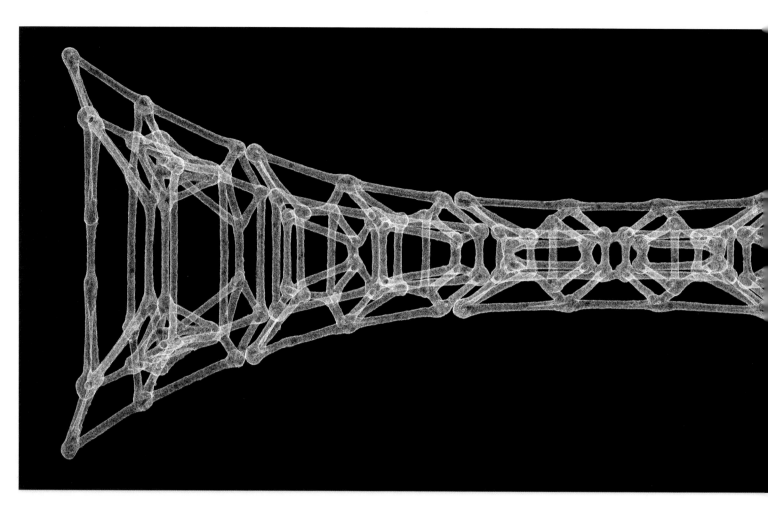

Yinan Xiao,
Virtually assembled bridge components,
Institute of Structural Design, Faculty
of Architecture, Civil Engineering and
Environmental Sciences,
TU Braunschweig, Germany,
2023

above: The 3D-scanned components were virtually assembled in a 3D software. Based on this virtual assembly, compression connectors were generated and 3D printed.

Yinan Xiao,
3D-printed bridge compression joint,
Institute of Structural Design, Faculty
of Architecture, Civil Engineering and
Environmental Sciences,
TU Braunschweig, Germany,
2023

opposite: Based on a virtual assembly of the 3D-scanned point cloud of the printed bridge components, Boolean operations were used to generate precise, individualised connectors that were 3D printed using flexible thermoplastic polyurethane.

Subsequently, the individual components were virtually assembled using the scan data in a 3D modelling software, leaving a 30-millimetre (1-inch) gap between the components. Based on this, precisely fitting compression connectors were created and 3D printed from a flexible thermoplastic polyurethane material. This digital-twinning process ensured the required accuracy for the compression connectors and met the high demands of the design. The individualised compression connectors were manually inserted to absorb stress concentrations when assembling the physical parts.

Reflections and Visions

The two case studies vividly illustrate the innovation potential of the Injection 3D Concrete Printing process, transforming concrete into lightweight lattice structures in an unprecedented departure from the conventional perception and tectonic logic of the material. The integration of VGS not only opens up unexplored avenues for highly material-efficient structures, but also introduces a paradigm of lightweight, resource-saving and cost-effective construction. Interpreting force vectors as intricate spatial print paths directly contributes to this efficiency, while the endlessly reusable suspension further enhances economic and ecological viability.

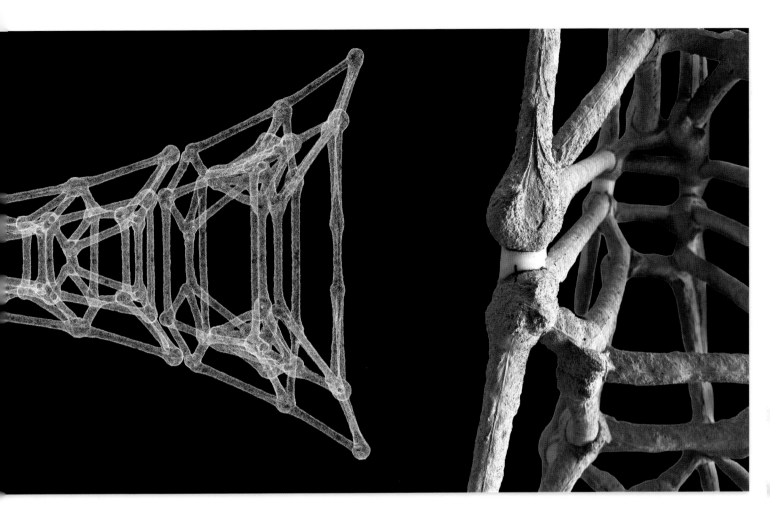

To advance the tectonic scope of the I3DCP process, investigations into reinforcement integration have been initiated. Co-extrusion of textile reinforcement struts and the encapsulation of reinforcement within the suspension have demonstrated significant promise. Incorporating reinforcement not only fortifies structures but also expands the formal spectrum to include tensile and bending configurations. This evolution necessitates novel approaches to connections capable of withstanding tension forces. Collaborative robotic processes have therefore been employed in scenarios where, for example, one robot positions steel connectors or reinforcement within the suspension while a second robot encases it in concrete. This expands the possibilities of structural design and connection methodologies, showcasing how technology can seamlessly integrate with architectural expression.

Adrian Forty's exploration underscores this dynamic interplay, emphasising how concrete continually reinvents itself through the inventive spirit of successive generations of architects and engineers, with Injection 3D Concrete Printing offering yet another perspective in reshaping the boundaries of form, structural efficiency and sustainability of concrete constructions. ᗪ

Notes
1. Adrian Forty, 'The Material Without a History', in Jean-Louis Cohen and Martin Moeller (eds), *Liquid Stone: New Architecture in Concrete*, Princeton Architectural Press (New York), 2006, pp 34–45.
2. Norman Hack *et al*, 'Injection 3D Concrete Printing: Basic Principles and Case Studies', *Materials* 13 (5), 2020, pp 1–17.
3. Dirk Lowke *et al*, 'Injection 3D Concrete Printing in a Carrier Liquid: Underlying Physics and Applications to Lightweight Space Frame Structures', *Cement and Concrete Composites* 124, 2021, pp 1–15.
4. Carl Culmann, *Die Graphische Statik*, Meyer & Zeller (Zurich), 1866.
5. Corentin Fivet and Denis Zastavni, 'Robert Maillart's Key Methods From the Salginatobel Bridge Design Process' [1928], *Journal of the International Association for Shell and Spatial Structures* 53 (1), 2012, pp 39–47.
6. Pierluigi D'Acunto *et al*, 'Vector-Based 3D Graphic Statics: A Framework for the Design of Spatial Structures Based on the Relation between Form and Forces', in David Hills and Stelios Kyriakides (eds), *International Journal of Solids and Structures* 167, 2019, pp 58–70.
7. Yinan Xiao *et al*, 'Injection 3D Concrete Printing Combined with Vector-Based 3D Graphic Statics', in Richard Buswell *et al*, *Proceedings of the Third RILEM International Conference on Concrete and Digital Fabrication*, Loughborough, 2022, pp 43–9.
8. Yinan Xiao *et al*, 'A Structure- and Fabrication-Informed Strategy for the Design of Lattice Structures with Injection 3D Concrete Printing', *Proceedings of the IASS Beijing Symposium: Next Generation Parametric Design*, Beijing, 2022, pp 1–12.

Text © 2024 John Wiley & Sons Ltd. Images: pp 30–31 Photo Norman Hack; pp 33(l), 35, 37–9 © Yinan Xiao; p 33(r) Martin Bond / Alamy; p 36 Photo Janna Vollrath

Fabio Amicarelli and Ena Lloret-Fritschi

Optimising Concrete Slabs with Paper Formworks

Fabrication and Material Aware
Architecture (FMAA) group,
Foldcast,
Academy of Architecture, Università
della Svizzera Italiana (USI),
Mendrisio, Switzerland,
2022-4

Design variations on scaled models. Paper is a low-cost and recyclable material, which can be three-dimensionally shaped according to predefined folding patterns. When used as material for formwork, this can enhance the design flexibility and variation of building elements, allowing the production of optimised structures with up to 50 per cent less concrete.

The constant demand for new buildings means concrete continues to contribute to a large proportion of global greenhouse gas emissions. Italian architect and researcher **Fabio Amicarelli**, and Assistant Professor in Architecture and Guest-Editor of this ⌂ **Ena Lloret-Fritschi**, discuss the Foldcast project being developed by the Fabrication and Material Aware Architecture (FMAA) group at the Università della Svizzera Italiana in Mendrisio. Foldcast combines software and digital machines to produce non-standard, recyclable, paper-based moulds for casting structurally efficient concrete elements.

Concrete is by far the most commonly used material in construction, making the cement industry responsible for around 6 per cent of CO_2 emissions worldwide.[1] With the global population predicted to rise to nearly 10 billion by 2050,[2] the demand for buildings is expected to double over the next 40 years.[3] Concrete being the most durable construction material, we can assume that it will continue to contribute to a significant proportion of the world's greenhouse gas emissions.[4] A reduction in the material's use is urgently needed across the planet, and there are new structural optimisation methods that could notably reduce its consumption and enhance construction efficiency. These methods are, however, often prohibitively expensive to implement, particularly because traditional construction systems lack adaptability and struggle to accommodate variation.

In the early 1990s, the introduction of digitally controlled machines in architecture promised to produce countless design variations of the same object at no extra cost.[5] However, this potential has not been fully realised in the construction industry. Currently, digital technologies are continuing to push its boundaries, aiming to bridge the gap between standardised building processes and customised design. Still, even after 30 years of digital construction, the fundamental question remains: can digital fabrication processes enhance how we create non-standard building elements, enabling design flexibility and fostering more sustainable construction processes?

Foldcast is a research project started in 2022 as part of the Fabrication and Material Aware Architecture (FMAA) group at the Academy of Architecture in Mendrisio, within the Università della Svizzera Italiana (USI). It combines computational tools and digital machines to produce non-standard formworks made of paper-based materials. This approach enables the production of structurally optimised building elements, using significantly less concrete while adopting low-cost and fully recyclable formworks. Foldcast thus illustrates that it is possible to significantly reduce the carbon footprint of concrete structural elements and make construction processes more efficient and sustainable.

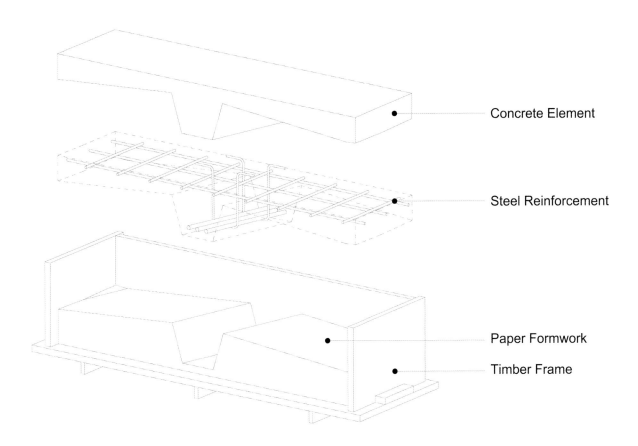

Concrete Element

Steel Reinforcement

Paper Formwork

Timber Frame

The Return of Structural Optimisation

Concrete is a resilient and versatile material that complies with nearly any desired shape when poured into a formwork. Formworks are typically constructed using wooden or metal panels and are essential for shaping uncured concrete and withstanding its pressure. Yet, to facilitate the use of such standard timber or metal formworks, slabs are often built with a uniform depth (flat slabs), using far more material than structurally necessary. Through such excessive use of concrete in structural elements, the construction sector has over the past decades favoured cost and time reduction over sustainability.

Posing an alternative, optimised systems like ribbed slabs employ about half the typical concrete volume by placing material only where strategically needed to ensure structural integrity. Ribbed slabs are flooring systems which consist of thin profiles and typically parallel reinforced concrete T-beams, and are usually produced by placing voids inside the formwork to reduce the bulk material of the slab and thus its weight. An early example is the

Fabrication and Material Aware Architecture (FMAA) group,
Foldcast,
Academy of Architecture, Università della Svizzera Italiana (USI), Mendrisio, Switzerland,
2022-4

Traditional construction systems use timber or metal panels to produce formworks. This results in expensive and labour-intensive processes when applied to optimised concrete elements. Foldcast employs digital technologies to design and manufacture custom formworks from fully recyclable paper, facilitating the creation of structurally optimised concrete structures. Shown here, an axonometry of paper formwork components.

Fabrication and Material Aware
Architecture (FMAA) group,
Foldcast,
Academy of Architecture, Università
della Svizzera Italiana (USI),
Mendrisio, Switzerland,
2022-4

below and opposite bottom: The paper formworks are designed with custom software, pre-cut using digital machines, and manually assembled into a formwork for concrete casting. This technology reduces production time and costs of custom formwork, optimises the use of materials and resources and allows an easy implementation in existing building processes.

Digital fabrication has the potential to produce flexible, non-standard formwork without notable additional costs, allowing customised concrete elements to be realised through industrial processes

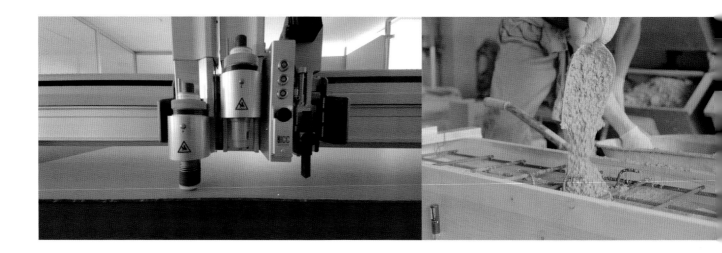

flooring system designed by Pier Luigi Nervi for the Gatti Wool Factory in Rome (1953), in which the curved ribs follow the principal bending moments of the floor.[6] Despite such benefits, ribbed slabs have still been sidelined, due to concerns about construction time and the cost of non-standard formwork made with timber or metal panels.[7]

Recently, digital design and fabrication processes have enabled the development of new methods to produce structural concrete elements with non-standard geometries, customised for individual requirements, without any notable influence on construction cost and time. These methods have thus made such structurally optimised concrete elements economically viable as a sustainable solution.

Improving Non-standard Concrete Formworks

Prefabrication involves the manufacturing of building components off site in an industrial setting. These elements are then transported to the construction site and assembled. Prefabricated building components have served as an alternative to on-site construction, improving efficiency, reducing waste, and ensuring higher-quality results in a safer environment. However, the modern-day paradigm of prefabrication is most suitable for the mass-production of standardised elements and is often associated with uniform and repetitive architecture. In this context, it is challenging to incorporate more complex, custom shapes such as those that are typically necessary for optimised structures. In the case of prefabricated concrete elements, the lack of flexibility

above: After the concrete hardening time, the wax-coated paper can be easily removed and fully recycled. A reusable paper-based supporting structure ensures that the formwork withstands the weight of the concrete.

is due to the high cost and time necessary to produce traditional formworks, which only become economically viable when reused several times. The inefficiency of producing variable structures often prevents designers from adopting non-standard systems.

Today, the most common method of creating non-standard formwork is to produce voids or inlays – usually made of expanded polystyrene (EPS) foam – that are then placed into standard scaffold systems and used for concrete casting. Although this method allows a wide range of forms to be created, the manufacturing process used to shape EPS generates substantial waste, as large quantities of material must be removed to produce the formwork.[8] Over the last two decades, academic researchers and industry have thus investigated how to improve non-standard formwork, using emerging technologies and testing new fabrication methods and materials. Digital fabrication has the potential to produce flexible, non-standard formwork without notable additional costs, allowing customised concrete elements to be realised through industrial processes. This paradigm shift opens up new design possibilities that embrace variation and differentiation over standardisation and repetition, while still taking advantage of the efficiency of industrial processes typical of prefabrication. In the bigger picture, the combination of non-standard formwork with off-site construction can thus reduce building costs and time, offer a wider range of architectural design options, and promote sustainable construction by saving materials and resources.

Fabrication and Material Aware
Architecture (FMAA) group,
Foldcast,
Academy of Architecture,
Università della Svizzera
Italiana (USI),
Mendrisio, Switzerland,
2022-4

right: The flexibility of paper formworks creates new design opportunities that value variation and uniqueness over standardisation. Supported by custom computational tools, architects and engineers can investigate a large variety of structural forms previously hard and expensive to build – such as this ceiling.

opposite: A 1:1-scale concrete slab element was produced in a research project involving researchers at the USI, University of Applied Sciences and Arts of Southern Switzerland (SUPSI) and Eastern Switzerland University of Applied Sciences (OST Rapperswil) along with industry partners Müller-Steinag Gruppe. From subsequent analysis and testing, it emerged that the prototype reduced the carbon footprint by 40 per cent overall compared to standard flat slabs made with timber formwork.

Paper Formworks for Concrete Elements

The Foldcast research project employs digital technologies to produce paper-based formworks, enhancing the structural optimisation of concrete elements. Leveraging paper, an affordable and widely available material with inherent ductility, the project emphasises sustainability by offering a fully recyclable solution. Through the precise folding of paper sheets based on designated patterns, intricate three-dimensional optimised geometries can be achieved with remarkable simplicity. The integration of custom computational tools and digital cutting machines is a key aspect of this approach, promising to significantly reduce fabrication time and costs associated with non-standard concrete formworks. In comparison to traditional custom timber or metal formwork, the paper-based alternative stands out for its sustainability – being lightweight, easily shaped, fully recyclable and effortlessly transportable. This innovation seeks to redefine construction practices by providing a more efficient and eco-friendly solution.

The fabrication process starts with the digital design of optimised concrete structural elements using custom computational tools that automatically generate a folding pattern to produce the corresponding paper formwork. The folding pattern, as a file format, is used to pre-cut flat sheets of paper with a digital machine. The pre-cut paper is then shipped to the prefabrication plant, where it is manually folded into a three-dimensional formwork inlay, placed into reusable scaffold systems and used for concrete casting. A waterproof coating and a paper-based supporting structure make the formwork resistant to the concrete pressure and humidity. After the curing, the formwork can be removed, and the paper can be reused or fully recycled. The final concrete elements produced with this method are designed to use up to 50 per cent less concrete than standard flat slabs, while fulfilling the same structural requirements. To ensure that the elements are code compliant, standard reinforcement steel and concrete mixes can be used.

The paper formwork technology has been developed through several preliminary tests that were intended to provide information to the custom computational tool that designs the paper components. A large-scale prototype of a structurally optimised concrete slab has been built as a collaboration between academic researchers at the USI, University of Applied Sciences

and Arts of Southern Switzerland (SUPSI) and Eastern Switzerland University of Applied Sciences (OST Rapperswil) and industry partners Müller-Steinag Gruppe. To further reduce the embodied carbon of the slab, the paper formwork was used in combination with a recycled concrete mix. Subsequent analysis showed that the prototype reduced the carbon footprint by 40 per cent overall compared to standard flat slabs made with timber formwork.

Design Variation and Sustainable Construction
Foldcast represents a material- and cost-effective strategy that significantly reduces the carbon footprint of concrete structures while allowing design flexibility and customisation. Enabled by custom computational tools, architects and engineers can explore a wide range of structural shapes that have until now been difficult to conceive and produce. Such tools empower designers with early-stage decisions that enhance structural and fabrication data, while making the production of non-standard formwork more efficient. The technology integrates innovative approaches into existing construction processes, aiming for easy implementation within the industry. Considering the number of buildings to be constructed in the near future, it is essential to pursue solutions that can streamline the construction processes, elevate the quality of architectural design, and contribute to moving the building industry towards a more sustainable future. 𝕯

Notes
1. UN Climate Change, 'Bigger Climate Action Emerging in Cement Industry', 26 October 2010: https://unfccc.int/news/bigger-climate-action-emerging-in-cement-industry.
2. UN Department of Economic and Social Affairs, *World Population Prospects: The 2017 Revision*, United Nations (New York), 2017: https://population.un.org/wpp/publications/files/wpp2017_keyfindings.pdf.
3. International Energy Agency, 'Global Building Sector CO_2 Emissions and Floor Area on the Net Zero Scenario, 2020–2050': www.iea.org/data-and-statistics/charts/global-buildings-sector-co2-emissions-and-floor-area-in-the-net-zero-scenario-2020-2050.
4. Architecture 2030, 'Why the Built Environment?': www.architecture2030.org/why-the-built-environment/.
5. Mario Carpo, *The Alphabet and the Algorithm*, MIT Press (Cambridge, MA), 2011, pp 98–9.
6. See Pier Luigi Nervi, *Costruire Correttamente*, 2nd edn, Ulrico Hoepli (Milan), 1965, pp 41–4.
7. Jaime Mata-Falcón et al, 'Digital Fabricated Ribbed Concrete Floor Slabs: A Sustainable Solution for Construction', *RILEM Technical Letters* 7, 2022, p 71.
8. Ena Lloret-Fritschi, 'Smart Dynamic Casting: A Digital Fabrication Method for Non-standard Concrete Structures', doctoral thesis, ETH Zurich, 2016, pp 33–4: www.research-collection.ethz.ch/handle/20.500.11850/123830.

Text © 2024 John Wiley & Sons Ltd. Images: pp 40, 43 © Fabrication and Material Aware Architecture (FMAA), Università della Svizzera italiana (USI); pp 44–6 © Università della Svizzera italiana (USI); p 47 © Alessandro Pio Gliaschera

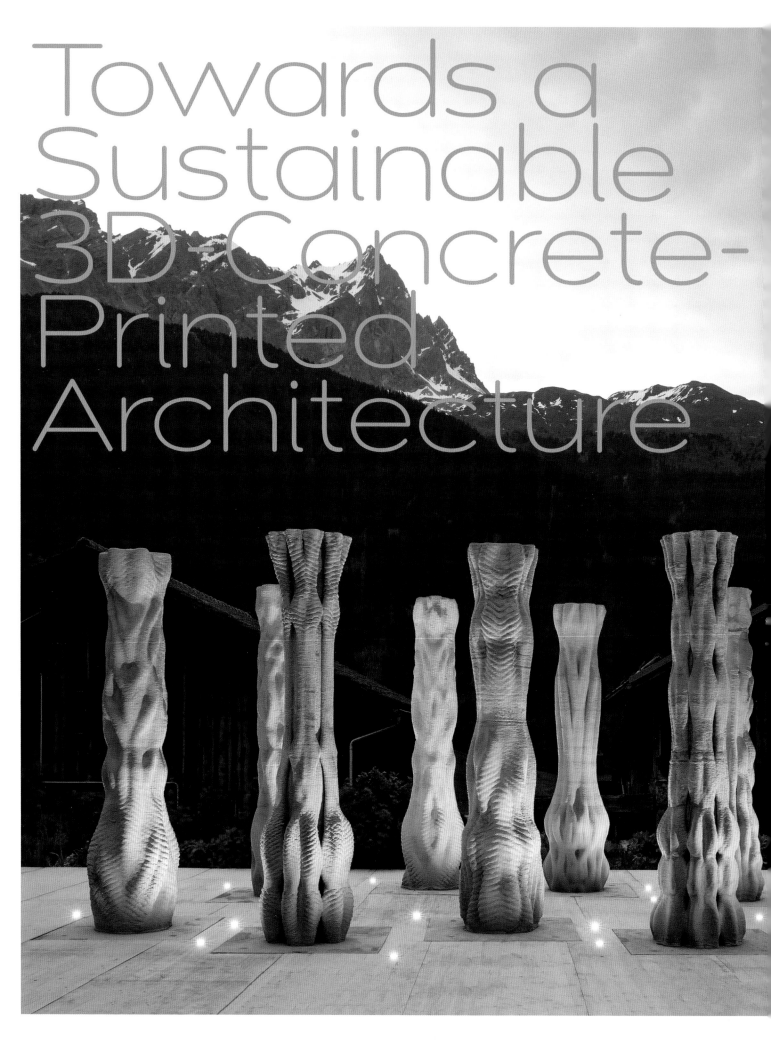

Towards a Sustainable 3D-Concrete-Printed Architecture

Ana Anton and Benjamin Dillenburger

Assemblies, Detailing and Ornamentation

Ana Anton and Benjamin Dillenburger /
Master of Advanced Studies in
Architecture and Digital Fabrication,
ETH Zurich,
Concrete Choreography,
Origen Festival of Culture,
Riom, Switzerland,
2019

opposite: The systematic design-to-fabrication strategy for the columns demonstrates that aesthetic diversity can be achieved at no additional cost by only changing the mid-scale and material-driven ornamentation.

left: The layer tectonics of every Concrete Choreography column consists of a cylindrical concrete cast core and an ornamented outer shell.

Benjamin Dillenburger
and Michael Hansmeyer,
Tor Alva,
Mulegns, Switzerland,
2024

The first and tallest tower with 3D-printed load-bearing columns, Tor Alva is an interdisciplinary research project developed in Switzerland through close collaboration between the Origen Foundation, ETH Zurich and local, national and international industry partners.

Given the long history of columns in the languages and lexicons of architecture, they are the ideal building elements with which to experiment in the pursuit of 3D-printed concrete. **Ana Anton and Benjamin Dillenburger** of the Chair for Digital Building Technologies at ETH Zurich reveal how new technologies can produce sinuous filigree structures that are unachievable with conventional casting approaches, and also offer more advantages.

Digital fabrication processes, including 3D concrete printing (3DCP), target resource efficiency in construction by strategically placing material only where necessary. A widely researched digital building technology, 3D concrete printing evolves by navigating process development, material performance, structural systems, construction methods and regulations. As a fabrication process, it employs a freshly mixed mortar based on Portland cement that undergoes an extrusion process to form a continuous filament. The filament is systematically deposited along a predefined path, layer by layer, enabling the construction of digitally designed objects. This incremental layering technique allows for shaping concrete without the need of a conventional formwork to support it as it sets.

Formwork-free concrete holds great sustainability potential for innovation in construction. 3DCP has undergone significant development over the past decades, progressing from the initial rotary analogue device for wall-printing in 1941[1] to the development of on-site layered extrusion deposition using a three-axes gantry system,[2] and further advancing with the utilisation of set-on-demand material systems[3] and sustainable formulations.[4]

The rationale for choosing 3DCP over conventional concreting methods has been based on considerations of geometric complexity, customisation, material efficiency and addressing labour shortages. Even though the number of 3DCP projects globally has grown exponentially in the past decade, the complexity of shapes has not seen a comparable evolution.

The adoption of 3DCP in construction is driven by the diverse strategies required for each use case, especially when customisation leads to higher production costs. For example, 3DCP can be executed directly on the construction site or in a controlled factory environment. On-site printing involves constructing a monolithic object positioned in its final location. In contrast, prefabrication produces several components in an optimal printability orientation, which are later transported and assembled on the construction site. Moreover, in construction-scale 3DCP, the choice between low-resolution printing and high-resolution printing has trade-offs. Low-resolution printing is faster and has larger aggregates but results in simpler geometries that require more material. High-resolution printing, even though it takes longer and uses materials with greater environmental impact, can create intricate shapes using less material overall. Furthermore, selecting a suitable system, such as funicular structures,[5] stay-in-place formwork[6] or the integration of steel reinforcement during material deposition[7] is instrumental in shaping the technology.

While formwork-free concrete construction is not unprecedented, determining a roadmap for optimal applications of 3DCP in construction poses critical questions. The challenges extend beyond the novelty of the technology to the intrinsic environmental impact of working with concrete, a material responsible for high CO_2 emissions. In a broader context, this highlights the need to define how this digital fabrication method can reshape building practices and its implications on concrete construction. Can 3DCP indeed contribute to lowering resource consumption in construction, including the materials from formwork, Portland cement and concrete itself? Additionally, are there other aspects of the technology that could further enhance sustainability in the construction sector?

While on-site printing with brick-size extrusion sections has certain environmental and productivity-related benefits,[8] current research by the Digital Building Technologies group at ETH Zurich explores an alternative model, focusing on the primary objectives of prefabrication, fine-resolution extrusion, geometric complexity, an optimised structural system, reinforcement integration, ornamentation and production quality. These are investigated in an interdisciplinary setup within the National Centre for Competence in Research – Digital Fabrication and have been implemented through two architectural-scale projects for the Origen Foundation. Through renovations, extensions and insertions of buildings for cultural activities along the Julier Pass valley in the Grisons, Switzerland, the Origen Foundation aims to infuse new vitality into a region experiencing a population decline.

Both projects showcase columns, the ideal architectural elements for exploring the design language and structural typologies enabled by 3DCP. They represent significant milestones in technology development, transitioning from using 3DCP as stay-in-place formwork to integrating the printed material structurally. Shifting from cylindrical-shaped columns to hollow-core branching columns is an evolution that equally illustrates progression in formal complexity and ornamentation. Effectively using 3D-printed concrete for structural purposes has a beneficial environmental impact as it ensures the material is not acting as a functionally inert layer inside the building. Instead, concrete is utilised for its load-bearing capacity while deposited in filigree patterns unachievable via conventional casting methods.

Concrete Choreography
The first collaboration with the Origen Foundation resulted in the production of a temporary installation consisting of nine 2.7-metre (8.9-foot) high 3DCP columns used as an outdoor stage design for the Origen Festival of Culture in 2019. The Concrete Choreography project was developed within a design studio for ETH Zurich Master of Advanced Studies in Architecture and Digital Fabrication 2018–19.[9] In this context, the team explored design methods for the column typology, establishing a suitable design language informed by the specific fabrication parameters of 3DCP.

The concise timeline and the design variations achieved by the Concrete Choreography columns demonstrated

Benjamin Dillenburger and
Michael Hansmeyer,
Tor Alva,
Mulegns, Switzerland,
2024

right: Located in a remote Swiss Alpine village, the tower is part of the cultural activities of the Origen Foundation to revive the area through architecture, art and technology.

opposite left: The tower consists of 41 prefabricated concrete components assembled with dry connections suitable for disassembly.

opposite right: The 3D-printed branching columns maintain the branching logic in the intersecting area, ensuring a continuous flow of forces.

that mass customisation for concrete construction is feasible. The columns feature an intricately ornamented exterior and a vertical hollow core filled with standard reinforced concrete. This design approach treats the columns as permanent formwork, where the 3D-printed material serves a purely decorative purpose. Despite this, the distinct and ornate design of the Concrete Choreography columns has allowed them to remain on-site for over five years, far exceeding the initial six-month duration planned for the project.

Tor Alva
Tor Alva is a 30-metre (98-foot) tall temporary tower in the remote village of Mulegns in the Grisons, designed in collaboration with architect Michael Hansmeyer for the Origen Foundation in 2024.[10] The structural design was developed in collaboration with engineer Jürg Conzett, from Conzett Bronzini Partner AG. The architectural function of this multistorey 3DCP structure is to host performances and art installations.

The project embraces sustainability across diverse socioeconomic, architectural and construction aspects. Alongside community engagement through cultural activities, the tower's construction involved collaborations with local construction companies. These partnerships introduced valuable digital expertise through computational design and robotic fabrication, facilitating knowledge transfer from academia to industry. The emphasis on modularity and built-for-disassembly strategies underpins circularity and reuse at the architectural level. Consequently, the tower will remain assembled on-site for a monitoring period of five years, and subsequently disassembled. After five years, the project will be reassembled in a different location.

The tower consists of 40 branching-column components ranging from two to four inclined branches. The top part of the structure is closed by a bespoke cupola that incorporates eight additional 3D-printed elements. Each branching-column component forms a structurally stiff triangle consisting of two precast segments at the top and bottom of a 3DCP branching element. The precast segments feature a conventional reinforcement. The 3DCP columns have shear reinforcement placed between the layers during printing. Vertical steel bars are also inserted post-printing into hollow channels connecting the precast segments with the 3DCP column. These channels are grouted to secure the structural bond between the longitudinal reinforcement, the printed material and the cast concrete. Additionally, the components of level four and the cupola are strengthened using post-tensioning, which involves inserting unbonded rods along the centre of each branch. For the 6-metre (20-foot) tall columns on level four, post-tensioning ensures their safe transportation on-site in a tilted position. For the cupola, which is designed as a compression-dominant structure, the columns rely solely on post-tensioning without any passive reinforcement.

These channels are grouted to secure the structural bond between the longitudinal reinforcement, the printed material and the cast concrete

Utilising branching columns is an excellent opportunity to showcase the importance of geometry in 3DCP

Benjamin Dillenburger and Michael Hansmeyer,
Tor Alva,
Mulegns, Switzerland,
2024

right: The branching columns are 3D-printed in the Robotic Fabrication Laboratory, using the custom setup developed through interdisciplinary research at ETH Zurich.

opposite: The material-driven ornamentation introduces variation among all columns within the same floor.

The 3DCP material of the Y column is structural and consists of three co-planar filaments. The outer filament creates the visible texture of the column. This ornamented outer layer serves an aesthetic function, introducing design variations among the columns and concealing the interface between the distinct 3DCP segments. The middle filament supports the inter-layer circular reinforcement bar, and the inner filament constructs the vertical channels. The middle and the inner filaments, together with the cast channel, create the structural body of the column. Despite the distinction between structure and ornament, there is no visible differentiation between the two. Both elements are seamlessly fused into the body of the 3DCP column, creating an integrated, durable and aesthetically cohesive design.

The 41 prefabricated structurally stable components are interconnected using dry assembly interfaces.

Utilising branching columns is an excellent opportunity to showcase the importance of geometry in 3DCP. This system enhances the structural behaviour of the tower, especially in responding to lateral forces such as wind or seismic loads. Hollow-core branching columns present fabrication challenges with conventional methods, especially at the intersection of branches where the rebar cage is notably complex. By deconstructing the rebar cage into vertical and horizontal steel bars supported by concrete layers, the fabrication of complex reinforced elements is simplified. This method highlights a novel approach in reinforced concrete, where the sequential assembly of concrete and steel results in optimal material configurations.

The progression from using 3DCP as permanent formwork in Concrete Choreography to incorporating the 3D-printed material for load-bearing purposes in Tor Alva was achieved by increasing the complexity of the column's cross-section. Demonstrating the benefits of using a procedural approach to print path design marks a change in future design methods. It emphasises the importance of designing both the exterior appearance and the internal structure of elements.

Designing for Sustainability
Both the Concrete Choreography and Tor Alva projects are exemplars of several years of interdisciplinary research at ETH Zurich aimed at using 3D-printed concrete directly as a load-bearing material. This unified vision was simultaneously researched in the fields of architecture, structural design and building material sciences, underscoring a powerful insight for the future of construction – that realising innovation and sustainability in buildings demands a comprehensive and collaborative strategy across diverse fields.

Clearly, the presented approach requires more research to further reduce the environmental footprint. However, a pivotal driver of innovation in construction lies in the seamless transfer of knowledge from academia to industry, and Tor Alva actively demonstrates this exchange through highly efficient design-to-fabrication protocols, infusing inspiration with the transformative impact of design. Incorporating smart logistics ensures that complexity is digitally mastered, enabling fabrication – similar to craftsmanship – to be localised adjacent to the construction site within a cutting-edge field factory.

Another dimension of this work engages with the idea that architectural design is responsible for adding value to the sociocultural sphere through spaces maintained for multiple generations. Concrete Choreography and Tor Alva demonstrate that 3DCP can facilitate a transition in concrete construction from mass production to mass customisation through automation.[11] In addition, constructive interactions with the client should not be underestimated. The urgency to produce fast iterations without compromising design integrity is a catalyst for innovation. These column explorations are materialised as affordable, structurally sound, thin-shelled hollow elements that incorporate ornamentation in a resilient material. As a result, digital fabrication with concrete re-establishes a connection between architecture and the inherent design variability of craftsmanship. ⌂

Notes

1. William E Urschel, 'Machine for Building Walls', United States Patent US2339892A, 1941.
2. Behrokh Khoshnevis, 'Automated Construction by Contour Crafting: Related Robotics and Information Technologies', *Automation in Construction* 13, 2004, pp 5–19.
3. Lex Reiter *et al*, 'Setting on Demand for Digital Concrete: Principles, Measurements, Chemistry and Validation', *Cement and Concrete Research* 132, 2020, 106047.
4. Federica Boscaro *et al*, 'Eco-Friendly, Set-on-Demand Digital Concrete', *3D Printing and Additive Manufacturing* 9 (1), 2022, pp 3–11.
5. Alessandro Dell'Endice *et al*, 'Structural Design and Engineering of Striatus, an Unreinforced 3D-Concrete-Printed Masonry Arch Bridge', *Engineering Structures* 292, 2023, p 116534.
6. Ana Anton *et al*, 'A 3D Concrete Printing Prefabrication Platform for Bespoke Columns', *Automation in Construction* 122, 2021, 103467.
7. Harald Kloft *et al*, 'Reinforcement Strategies for 3D-Concrete-Printing', *Civil Engineering Design* 2, 2020, pp 131–9.
8. Viktor Mechtcherine *et al*, 'Large-Scale Digital Concrete Construction: CONPrint3D Concept for On-Site, Monolithic 3D-Printing', *Automation in Construction* 107, 2019, 102933.
9. Ana Anton *et al*, 'Concrete Choreography: Prefabrication of 3D-Printed Columns', in Jane Burry *et al*, *Fabricate* 2020: *Making Resilient Architecture*, UCL Press (London), 2020, pp 286–93.
10. Ana Anton *et al*, 'Tor Alva: A 3D Concrete Printed Tower', in Phil Ayres *et al*, *Fabricate 2024: Creating Resourceful Futures*, UCL Press (London), 2024, pp 252–9.
11. This work was funded by the Swiss National Science Foundation, within the National Centre for Competence in Research in Digital Fabrication (project number 51NF40-141853).

Text © 2024 John Wiley & Sons Ltd. Images: p 48 © Benjamin Hofer, Digital Building Technologies, ETH Zurich; p 49 © Keerthana Udaykumar, Digital Building Technologies, ETH Zurich; pp 50, 52, 53(l) © Benjamin Dillenburger and Michael Hansmeyer, Digital Building Technologies, ETH Zurich; p 53(r) © Nijat Mahamaliyev, Digital Building Technologies, ETH Zurich; p 54 © Ana Anton, Digital Building Technologies, ETH Zurich; p 55 © Michael Hansmeyer, Digital Building Technologies, ETH Zurich

David Jenny, Fabio Gramazio and Matthias Kohler

REIMAGINING EARTHEN MATERIALS

THE NEW ERA OF SUSTAINABLE AND DIGITAL CONSTRUCTION

Gramazio Kohler Research, ETH Zurich,
Clay Rotunda, SE MusicLab,
Bern, Switzerland,
2021

left: Finalised structure in its context after completion of the interior auditorium and addition of the lightweight roof structure. The cylindrical wall is a material expression of its construction process, visibly showing traces of its segmentation and building sequence.

above: Close-up of the structure highlighting the unique material expression that originates from the plastic deformations of the soft clay cylinders when pressed in place during the fabrication process. A novel aesthetic at the interplay of digital control and material behaviour.

EARTHEN BUILDINGS ARE UBIQUITOUS THROUGHOUT MOST OF THE WORLD, AND ARE STILL BEING CONSTRUCTED TODAY. HOWEVER, IN THE WAKE OF OUR GREATER UNDERSTANDING OF CLIMATE CHANGE, THIS HISTORIC ARCHITECTURAL MATERIAL IS BEING SEEN AND USED IN A NEW LIGHT, AUGMENTED BY 21ST-CENTURY ROBOTIC DEXTERITY. GUEST-EDITOR OF THIS △ **DAVID JENNY**, AND **FABIO GRAMAZIO AND MATTHIAS KOHLER**, CO-FOUNDERS OF THE ARCHITECTURE AND DESIGN GROUP GRAMAZIO KOHLER RESEARCH AT ETH ZURICH, DISCUSS SOME OF THEIR RECENT ACHIEVEMENTS IN THIS RESPECT.

Clay and other earth-based substances are among the most ancient natural building materials and are still widely used today. Approximately one-tenth of the global population is estimated to live and work in earthen buildings, most of which are simple constructions using traditional vernacular techniques.[1] However, during the 20th century and with the advancing industrialisation of construction processes, the significance of clay as a building material became marginalised. Most notably, unfired clay has gradually been replaced by more processed materials. In the late 19th century, as part of the Second Industrial Revolution, the introduction of industrialised mass-production of fired-clay bricks and later fired limestone for cement production – the basic ingredient of reinforced concrete – marked a significant shift. Although their production and application depend on energy- and emission-intensive processes and result in not easily recyclable or sustainable constructions, these materials have been widely employed over the last century as they overperform clay in their structural capability and durability. Consequently, earthen construction techniques have seen little technological development. Techniques like rammed-earth construction, which involves mixing and compacting earth between formworks, are labour intensive and time consuming. Such methods remain economically viable mostly in regions with low labour costs and limited technological development. As a result, despite its abundance and sustainability, unfired clay plays a very limited role in today's industrialised construction processes.

Old Materials, New Constructions

The building sector, a major contributor to emissions and waste, urgently requires disruptive and transformative changes. Sustainability is a core challenge and main driving force in new construction projects. The need to reduce the amount of embodied energy in building materials and to integrate them into local and circular economies opens up exciting opportunities to revisit the potential of unfired clay as a locally available, fully reusable and emission-free building material.[2] In this context, the project presented here brings forward a radically new approach to earthen construction. The Clay Rotunda, realised by Gramazio Kohler Research at ETH Zurich, is a unique, free-standing cylindrical structure that constitutes the external, soundproof shell of the SE MusicLab, a high-fidelity music auditorium built inside the newly refurbished Gurten Brewery in Bern, Switzerland. It combines the power of computational design and the precision of robotic fabrication with clay's natural and malleable material behaviour. The project demonstrates how complex material behaviour in combination with digital technologies reshapes the way we build; we are moving from rigid control over industrially produced parts to a more adaptive building process making use of non-homogeneous materials. This shift paves the way for a novel aesthetic that bridges the natural world and digital innovation.

As the first permanent implementation of a novel earth-based robotic fabrication process at full architectural scale, the Clay Rotunda addresses the critical issue of reducing material consumption and associated emissions in construction by combining clay – a sustainable zero-waste building material – with cutting-edge computational design and robotic fabrication techniques. Structurally optimised, the Clay Rotunda features an ultra-thin wall with a diameter of 11 metres (36 feet), standing 5 metres (16 feet) tall, yet only 15 centimetres (6 inches) thick, and is built entirely out of unreinforced and unfired clay. It was implemented using a bespoke mobile robotic platform known as the in-situ fabricator,[3] which over 50 days meticulously assembled more than 30,000 soft clay bricks on-site, demonstrating the feasibility and precision of this new construction method. The realisation of the project is the fruit of a radically interdisciplinary collaboration involving partners from both industry and academia.[4] As such, it can be understood as a ground-breaking architectural experiment, acting as a catalyst for knowledge transfer between research and practical application in the industry.

Design, Structure and Segmentation

The design of the Clay Rotunda responds directly to the spatial requirements of the music auditorium and the constraints imposed by the surrounding environment, including ceiling height and adjacent walls. Minimising the wall thickness presented a significant structural challenge. The design approach not only optimised space usage within the Clay Rotunda, but most importantly significantly reduced production time. Achieving such slender dimensions while ensuring the stability and durability of the structure required innovative engineering solutions. While earthen constructions typically demand wide cross-sections as they mainly work under compressive loads, the extreme slenderness of the wall is made possible by its undulating base. This increases the footprint and thus stabilises the structure and prevents buckling. The wave amplitude decreases along a parabolic function over the structure's height, culminating in a circular form at the top. Overall, this results in a double-curved shell that elegantly marries architectural aesthetics with structural performance.

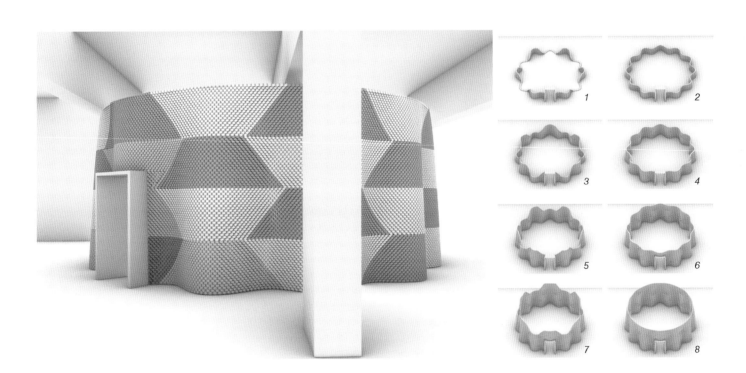

Gramazio Kohler Research, ETH Zurich,
Clay Rotunda,
2020
Overall design and segmentation diagram (left), showing the horizontal and vertical construction sequence (right), each construction step consisting, respectively, of five and six trapezoidal segments and relating to approximately one week of building time. Individual segments are built separately to allow for the initial drying and shrinkage to take place before connecting subsequent segments.

Gramazio Kohler Research, ETH Zurich,
Remote Material Deposition installation,
Sitterwerk,
St Gallen, Switzerland,
2014

right: Through the controlled throwing of material, a small-sized robotic system exceeds its workspace and deposits material at a remote location, building larger architectural structures over distance. While digitally controlled and informed through a 3D-scanning feedback process, the ballistic behaviour of the individual cylinders leads to a stochastic distribution within the aggregated structure.

Gramazio Kohler Research, ETH Zurich,
Clay Rotunda,
2020

above and opposite: Diagram showing the gradual orientation (above left) of individual cylinders within each segment from 0° at the centre to 30° at the edge, and their construction sequence (opposite right), depending on the orientation of the trapezoid. The precise positioning of the robotic arm as it presses onto the seams on the inclined edges ensures proper bonding and integral structural behaviour among the segments.

Constructive System

A sophisticated computational model was essential for designing the Clay Rotunda's thin shell. Development of the model considered the calculations of the engineers, the specific properties of the clay mix, as well as the constraints of the fabrication process. It played a critical role in generating the overall geometry and in segmenting the structure into manageable elements by calculating the position and orientation of each of the more than 30,000 clay cylinders. Accounting for the limited reach of the mobile robotic platform and the clay's shrinking and drying behaviour, a strategy for horizontal and vertical segmentation into matching trapezoids was implemented. Within each segment, this allowed for computation of the building sequence of the individual clay cylinders. Given the plasticity of the material, the order and method of placing and pressing were crucial to achieving the desired geometric outcome and directly influenced its material expression. An important feature of this process

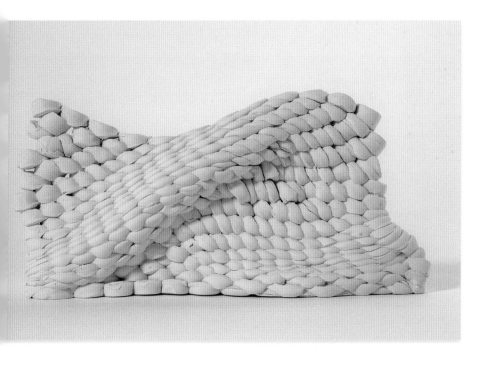

Gramazio Kohler Research, ETH Zurich,
Rapid Clay Formations,
2018

left: Rapid Clay Formations was first investigated at small scale, where a collaborative robotic arm prepares projectiles of different sizes, positions them, and presses them into their respective shapes by linear pneumatic actuation. The precise control of forces and positions applied to the material allows the design and build of highly differentiated and articulated clay structures.

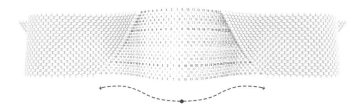
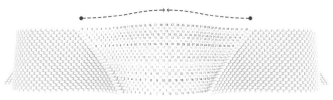

was the gradual orientation change from the centre to the edge of each trapezoid. This was particularly relevant for ensuring consistency in the material distribution and proper bonding between the different segments. Finally, all necessary fabrication data to guide the robotic placing and pressing process was generated through the same computational model, facilitating the realisation of the bespoke structure at high geometric complexity.

The innovative concept of robotic clay aggregation originates from the Remote Material Deposition project realised by Gramazio Kohler Research in 2014. This pioneering installation was conceived as a playful yet radical departure from traditional 3D-printing methods. It introduced the idea of depositing materials from a distance by catapulting portions through the air with a robotic arm. The inherent unpredictability in the precise landing position and orientation of the material was offset by a real-time feedback mechanism. This system continually adjusted the construction model based on the 'as-built' state, captured by a LiDAR scanner. While this approach had its practical limitations, it facilitated the creation of architectural structures with distinctive morphologies and aesthetics. The beauty and uniqueness of these structures are the direct result of the dynamic and adaptive fabrication process employed in their creation.[5]

This innovative idea was then embraced and evolved within the Master of Advanced Studies in Architecture and Digital Fabrication at ETH Zurich. Through multiple iterations and at varying scales, in the Rapid Clay Formations investigation the focus shifted towards a more controlled deposition method whereby malleable clay cylinders are picked and pressed into a precise bond by a robotic arm. The Clay Rotunda stands as the inaugural permanent architectural application of this fabrication process.[6]

Gramazio Kohler Research, ETH Zurich,
Clay Rotunda,
SE MusicLab,
Bern, Switzerland,
2021

right: A robotic manipulator precisely positions a soft clay cylinder in a bespoke angle right before pressing. The soft brick is held by two retractable pneumatic picks and released shortly before the robotic arm presses it into its final position.

opposite: The Clay Rotunda structure right after construction was completed. Lighter and darker areas indicate the progress of the drying process. The structurally optimised double-curved geometry of the wall enhances its stability and supports a controlled shrinkage.

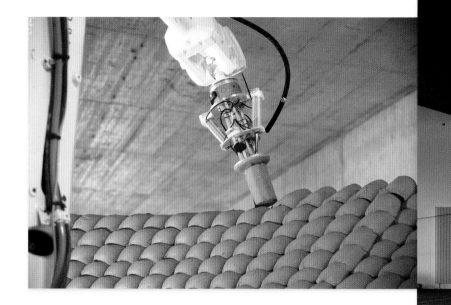

Material System and Robotic Fabrication

As the success of the fabrication process heavily depends on the behaviour and performance of the material, scaling up to full architectural scale required the refinement of the material mix. Different combinations of locally sourced clay, sand, grit and water were tested iteratively. Faced with the challenge of avoiding cementitious additives, the key was to find an optimal balance. This involved achieving sufficient plasticity for the fabrication process and material bonding, maximising compressive strength for structural integrity, and minimising material shrinkage to prevent deformations and cracks. The resulting mix, comprising 40 per cent clay, 45 per cent sand and 15 per cent grit with a 1/16th proportion of water, is extruded to create so-called 'soft bricks'. These malleable clay cylinders, measuring 90 millimetres (3.5 inches) in diameter and 150 millimetres (6 inches) in height, form the fundamental building blocks of the structure.

The constructive system employed in the project is a contemporary reinterpretation of traditional earthen construction techniques, like cob walls. To accommodate the dimension of the structure, the bespoke robotic platform is relocated for each segment.[7] In the building process, each soft brick is robotically picked using a custom gripper, precisely oriented and then pressed against the others in a quasi-layered fashion. Leveraging on the malleability of the wet material, the cylinders are compressed to about 60 per cent of their original height, thereby creating a strong connection to the previously deposited cylinders through surface cohesion and geometric interlocking. The process generates a unique material expression originating from the plastic deformation of the soft bricks. It defines an entirely novel aesthetic, emerging from the interplay between digital control and the intricate behaviour of a non-homogeneous material.

The Art of Building the Clay Rotunda

Utilising a non-standard material with somewhat unpredictable behaviour calls for a paradigm shift in the building process. As the structure dries and water evaporates, material shrinkage causes cracks and deformations until it has completely hardened. While the cracks can be easily repaired using the same clay mix, managing the deformations during the drying process requires more sophisticated monitoring and adaptation strategies. Instead of relying solely on the precision of the fabrication process, regular acquisition of the as-built geometry through 3D scanning is employed. This approach enables the building team to observe deformation as it happens and to respond to it. The strategy is not to force the material to conform to the design; rather, the computational model is continuously updated based on real-time data. This allows for gradual adjustments in the placement of clay cylinders to align with the actual conditions on-site. Acknowledging that deformations are inevitable and cannot be precisely simulated or predicted echoes the traditional working methods of medieval masonry, where adaptation and on-site decision-making were integral to the construction process.[8] Simultaneously, this approach necessitates proactive interdisciplinary collaboration among all stakeholders involved in the design process. This early engagement is crucial for devising robust strategies to effectively mitigate potential risks and to collectively share liabilities.

The Clay Rotunda is an example of how state-of-the-art digital design and fabrication techniques can make use of traditional material and construction knowledge to radically challenge our current approach to design and building. It demonstrates the potential to reduce material consumption by leveraging on complex geometries, and highlights the viability of reverting to less processed, zero-emission and waste-free natural building materials in the construction of our built environments. While the increased level of control over construction processes allows building complex structures beyond traditional possibilities, it also invites us to question some of the core values ingrained in our building culture. The Clay Rotunda is a tangible experiment that challenges our aesthetic sensibilities while prompting us to reconsider our definitions of performance and quality in architecture. It emphasises the crucial role of maintenance and care as integral components of a building's life cycle. This structure embodies radical sustainability, continuing the rich legacy of earthen construction. With proper maintenance and care, it is designed to endure indefinitely or until such time as it can be completely returned to nature. ⌂

Notes
1. See the up-to-date study: Alastair TM Marsh and Yask Kulshreshtha, 'The State of Earthen Housing Worldwide: How Development Affects Attitudes and Adoption', *Building Research and Information* 50 (5), 2022, pp 485–501.
2. See, for instance, Roger Boltshauser, Cyril Veillon and Nadja Maillard (eds), *Pisé – Rammed Earth: Tradition and Potential*, Triest Verlag (Zurich), 2019.
3. Markus Giftthaler *et al*, *Mobile Robotic Fabrication at 1:1 Scale: the In situ Fabricator in Construction Robotics*, Springer Nature (Heidelberg), 2017.
4. For full project credits visit: gramaziokohler.arch.ethz.ch/web/projekte/e/0/0/0/430.html
5. Kathrin Dörfler *et al*, 'Remote Material Deposition', in Maria Voyatzaki (ed), *What's the Matter? Materiality and Materialism at the Age of Computation*, ENHSA (Barcelona), 2014, pp 361–77.
6. David Jenny *et al*, 'A Pedagogy of Digital Materiality: Integrated Design and Robotic Fabrication Projects of the Master of Advanced Studies in Architecture and Digital Fabrication', in Marie Frier Hvejsel (ed), *Architecture, Structures and Construction* 2 (4), 2022, pp 649–60.
7. Selen Ercan *et al*, 'Automated Localization of a Mobile Construction Robot with an External Measurement Device', in *Proceedings of the 36th International Symposium on Automation and Robotics in Construction*, Banff, 2019, pp 929–36.
8. David Jenny, 'Daten in Material giessen: Digital-analoge Konstruktionen', in Patric Furrer, Andreas Jud and Stefan Kurath (eds), *Digitalisierung und Architektur in Lehre und Praxis*, Triest Verlag (Zurich), 2022, pp 87–95.

Text © 2024 John Wiley & Sons Ltd. Images: pp 56–7 © Gramazio Kohler Research, ETH Zurich, photos Michael Lyrenmann; pp 59, 60–61(b), 62–3 © Gramazio Kohler Research, ETH Zurich; p 60(t) © Gramazio Kohler Research, ETH Zurich, photo Yves Roth; p 61(t) © Gramazio Kohler Research, ETH Zurich, photo David Jenny

Jelle Feringa

Scalable Equals Sustainable

FRAGMENTATION

AMALGAMATION

The Infrastructural Imperative of Earthen Construction

Terrestrial,
Rock cycle versus
accelerated rock cycle,
2023
The robotic shot-earth 3D printing SE3DP method is analogous to an accelerated geological rock cycle, applying pressure not gradually over time but in an instant, transforming earth, gravel and clay into architectural forms. This construction method embodies a move away from extractive and unsustainable building practices, while taking robotic and additive manufacturing methods to the scale of infrastructure.

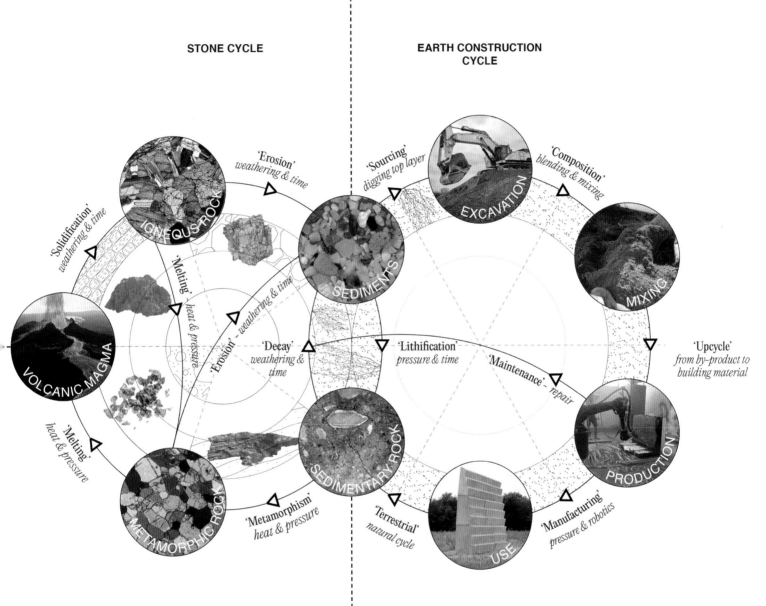

Building with earth seems a sensible ambition in our troubled climatic times. A fundamental problem, though, is being able to make larger, more complex structures from earth without jettisoning its unique sustainable and artisan qualities at the domestic and sub-domestic scales. Architecture and robotics specialist and co-founder of EZCT Architecture & Design Research **Jelle Feringa** considers the concepts and issues involved.

Terrestrial,
Field test of the printed earthen acoustic barrier,
Eindhoven, The Netherlands,
June 2023
left: Field test of Terrestrial's earthen acoustic barrier, which represents a first in demonstrating the novel SE3DP process.

Terrestrial,
Printed earthen acoustic barrier,
Rotterdam, The Netherlands,
2022
opposite: Terrestrial's ambition is the notion of an earthen acoustic barrier that transcends its utilitarian guise. The sinusoidal shape of the wall simultaneously augments the wall's strength and increases its acoustic insulation capacity while also lending it a sculptural elegance.

Buildings, too, are children of Earth and Sun.
— Frank Lloyd Wright, 1932[1]

While the urgency of finding ways to make architecture non-extractive is broadly understood,[2] there is a considerable lack of construction methods to meet this demand. Given the abundance, omnipresence and consequent affordability of earth, is its potential to address the worldwide resource crisis being held back by the archaic method of 'ramming' that is almost eponymous with earthen construction? This artisanal approach harbours a certain melancholy when it comes to addressing global construction challenges with earthen materials, as it operates on a scale too small to effect significant change. That is the dilemma of scaling up earthen construction: how to make a substantial impact without discarding the core values of sustainability and ecology. A responsible approach towards construction is scalable; scalable equals sustainable.

Terrestrial is a startup that explores the merits of robotic fabrication and innovates novel and scalable approaches to building with earth. The company's pilot project is for an earthen acoustic barrier, working towards the automation of earthen construction so that it can be used at an infrastructural scale.

The Accelerated Rock Cycle

Building with earth presents an ontological paradox. Within the terrestrial biosphere, the white noise of entropy is the universe's melancholic hymn of disintegration, dissolution and the inevitable gravitation towards chaos. The weathered offspring of geological time embodies a contradiction. Clay is the sediment that has endured the most prolonged exposure to fracturing, grounding, sharding and shattering, splintering, scraping and the scattering of entropic forces. Clay particles have been smithered for aeons – far longer than their juvenile counterparts, gravel and sand – and in this process of degradation, clay has acquired its potent force of cohesion in resistance to the process that wielded it. To build with earth is to reconstitute weathered sediment, to reverse entropy. The process of robotic shot-earth 3D printing (SE3DP) represents an accelerated rock cycle.

The irony that clay – the minutiae of the geological world – embodies is that while weathered to its finest, it eventually transmutes into a binding agent. Clay epitomises negentropy, an ordering force that opposes the decay and disorder of the world. Analogous to how cement constitutes concrete, that task falls to clay in earthen construction. A profound parallel between SE3DP and the rock cycle emerges – in SE3DP, the residue of one rock cycle provides the means to reconstitute. In the natural world, pressure slowly compacts clay and other sediments, binding these into sedimentary rock. Over time and under even greater pressures and temperatures, rocks can transform further into the metamorphic rocks that make up much of our Earth's crust. SE3DP presents a poetic parallel of the natural rock cycle, using pressure not in geological steps but momentarily, rapidly compacting earth, gravel and clay into architectural structures. Building with earthen materials implies a rejection of the extractive and unsustainable constructions.

> While considerable research focuses on improving aspects of materiality in earthen construction, there needs to be more innovation in the processing of earthen materials in order for a scalable, industrial approach to be established

Scalable Is Sustainable

While considerable research focuses on improving aspects of materiality in earthen construction, there needs to be more innovation in the processing of earthen materials in order for a scalable, industrial approach to be established. The challenge is that artisanal handling and processing overshadow raw material acquisition. There is false sentimentality in craftsmanship; the artisanal take is often associated with ecological approaches to construction. The lack of an industrial approach to earthen construction is at odds with the implicit egalitarian notion that earthen construction embodies – the Earth, and therefore earth, belongs to anyone and everyone. That aspiration of universality echoes the statement by American artist and leading figure in the Pop Art movement Andy Warhol that 'A Coke is a Coke, and no amount of money can get you a better Coke than the one the bum on the corner is drinking. All the Cokes are the same and all the Cokes are good.'[3] From this perspective, vernacular approaches can sometimes be counterproductive. While suggestive of a humanistic approach to construction, they cost so much more than conventional methods that are antagonistic to the suggestion of universality that earthen architecture implies.

Nick Srnicek and Alex Williams's book *Inventing the Future* (2015) summarises the chagrin that has resulted from the shift from the cheekiness and optimism of the Pop Art era to vexing populism and gloomy post-capitalism: 'Modernisation has come to signify simply some dread combination of privatisation, heightened exploitation, rising inequality and inept managerialism. Likewise, notions of the future tend to revolve around ideas of ecological apocalypse, the dismantling of the welfare state, or corporate-led dystopia, rather than anything bearing the mark of utopia or universal emancipation. For many, therefore, modernity is simply a cultural expression of capitalism.'[4]

In the audacious narrative of architectural evolution, where concrete once reigned as the undisputed structural heroin(e) – its very essence coursing through the veins of Modernism, offering a high from which we dared not descend – we now stand at a precipice, gazing into the terra firma of earthen architecture with a resolve to go cold turkey. This shift is not merely a retreat from the intoxicating embrace of concrete but a radical, deliberate recalibration towards the elemental, the primal, the sustainable.

Earthen architecture juxtaposes the Late Modern ethos of universal access represented by mass consumption with the resurgence of artisanal, vernacular approaches in retort to the disillusionment with modernity. Deeply rooted in local materials and craftsmanship, it in theory embodies a promise of architectural universalism – accessible and part of human heritage across cultures and geographies. Essentially, it should be as universally accessible as a bottle of Coke.

As the reaction against modernity equates it with capitalism and its ills, the return to artisanry and craft in architecture, as conveyed in earthen buildings, paradoxically moves away from an architecture that is accessible. While aiming to reject homogenisation and material exploitation, this shift towards vernacular architecture inadvertently creates exclusivity due to the practical and economic constraints of the current building practice.

The haughtiness of craft lies in the juxtaposition of access and the reality of cost and practicality. It is ironic that the emerging consumer culture of the Pop Art era prompts the realisation that few things have since become more economically divisive than (having access to) architecture. Neo-vernacular architecture and earthen construction's current dependency on intense labour and artisanry hinders its accessibility. The German Association for Building with Earth (Dachverband Lehm) indicates a worktime of 8 to 12 hours for a cubic metre of rammed-earth wall,[5] which results in a price point that stifles impact.[6]

In its day, Warhol's Coke bottle represented a democratising element in modern culture – a universal experience accessible to all, irrespective of status or wealth. At the current price point, earthen architecture embodies a tension that amounts to hypocrisy – through the use of earth, this universality is suggested but not effectuated. In addressing the challenges of the housing market, and those of climate change, few things are as pressing as the development of an accessible and sustainable approach to construction.

The considerable cost of a non-industrial approach to sustainable construction surrenders ambition. To align earthen construction with its democratic appeal is to engage in industrial research. The architect is not just a builder of structures, but a responsible steward of resources. To quote French Neoclassicist architect Claude-Nicolas Ledoux: 'It is for the architect to oversee the principle; he can activate the resources of industry husband its products and avoid costly upkeep he can augment the treasury by means of the prodigal devices of art.'[7]

Martin Rauch's pioneering work has breathed new relevance into earthen construction. With his company Lehm Ton Erde (meaning 'Loam Clay Earth') he has spearheaded the development of a mechanised approach to building with rammed earth. Working on the acclaimed Ricola Kräuterzentrum in Laufen, Switzerland (2014) – a herb processing facility for herbal cough drops and tea manufacturer Ricola, designed by Herzog & de Meuron – Rauch highlighted the project's necessity for efficiency, stating: 'In a country with Europe's highest hourly wage rates, this

Lehm Ton Erde,
Mechanised method for rammed-earth construction,
Schlins, Austria,
2022

Martin Rauch and his company Lehm Ton Erde pioneered this construction method, which was a critical step towards the realisation of Herzog & de Meuron's Ricola Kräuterzentrum in Laufen, Switzerland (2014).

project could only be realised through rationalisation. Ricola gave us the order for the machine development before they even knew if they would actually build it.'[8] The conjunction of the builder/machine builder recalls the Italian Renaissance architect Filippo Brunelleschi, who secured the world's first recorded patent, granted in 1421 for an improved method of transporting goods up and down the river Arno in Florence.[9]

Where Lehm Ton Erde is mechanising an approach to the established practice of rammed-earth construction, Terrestrial is rethinking earthen construction from the ground up. For the ramming of earth, formwork is required to ensure that the final density, strength and durability are met. These take up to 60 per cent of site operations.[10] Rather than compacting earth in a shuttering mould, in SE3DP the material is printed at a high volume and compacted with pneumatic pressure. What makes the approach so much more productive is the use of high-pressure air to both convey the soil mixture to the formwork and provide the force of impact, which contributes to wall strength and durability. Due to the high-impact velocity of the material stream, there is no formation of cold joints even when layers are built up weeks apart from one another.

Unlike 'conventional' 3D concrete printing, SE3DP is essentially mechanical compaction, not a chemical process. A chemical reaction is sensitive to environmental conditions – humidity and temperature – and thus challenging to operate in the field, while SE3DP is a robust process that is largely insusceptible to environmental conditions. Above all, with the SE3DP print method, a high-deposition volume is attainable to print several cubic metres per hour – an output level that is crucial when working in infrastructural applications.

Terrestrial is working towards an integrated approach that seamlessly combines product development with production methodology. This strategy mirrors the efforts of Aectual, a firm known for its innovative work in architectural design and fabrication, where the creation of architectural products and the development of 3D-printing methods occur simultaneously.

The next step in Terrestrial's development of earthen acoustic barriers, following on from an earlier field test, is the production of a 130-metre (426-foot) prototype. Infrastructure is uncharted territory, while synonymous with earth-moving and landscape architecture. Where architectural projects are more incidental, high-volume projects such as the over 55 kilometres (34 miles) of acoustics barriers that will be built before 2030 – representing over 48,000 tons of CO_2 equivalent – can provide the bedrock for more drawn-out technological developments. Similarly, the Dutch Ministry of Infrastructure and Water Management will install a comparable body of acoustic barriers in the same timeframe.

Aside from the rational motivations and the gravitas of a project poised to redefine the landscape, Terrestrial's work embodies a haunting architectural reverie: the notion of an earthen acoustic barrier that transcends its utilitarian guise, to be envisioned as a monumental act of land art. 🟥

Notes
1. Frank Lloyd Wright, *Frank Lloyd Wright: An Autobiography*, Longmans, Green and Company (Toronto), 1932, p 149.
2. See Space Caviar and V-A-C Foundation (eds), *Non-Extractive Architecture: On Designing Without Depletion*, V-A-C Press (Moscow), 2021.
3. Andy Warhol, *The Philosophy of Andy Warhol: From A to B and Back Again*, Harcourt Brace Jovanovich (New York), 1975, p 100.
4. Nick Srnicek and Alex Williams, *Inventing the Future: Postcapitalism and a World Without Work*, Verso Books (New York), 2015, p 48.
5. Horst Schroeder, *Sustainable Building with Earth*, Springer International Publishing (Vienna), 2016, p 294.
6. Tobias Helmersson, *From the Ground Up: Research on Rammed Earth and Timber for a Residential Building*, Master's thesis, Chalmers University of Technology (Gothenburg), 2002, p 82.
7. Anthony Vidler, *Claude-Nicolas Ledoux: Architecture and Social Reform at the End of the Ancien Régime*, MIT Press (Cambridge, MA), 1990, p 41.
8. Martin Rauch, *Martin Rauch: Stampflehmbau auf dem Weg zum postfossilen Bauen*, Österreichischen Lehmbautagung, Architekturzentrum Wien (Vienna), 2023: https://www.youtube.com/watch?v=nwc1CwOEW-g.
9. Frank D Prager and Gustina Scaglia, *Brunelleschi: Studies of His Technology and Inventions*, Dover Architecture (Mineola, NY), 1970, p 124.
10. Domenico Gallipoli *et al*, 'A Geotechnical Perspective of Raw Earth Building', *Acta Geotechnica* 12 (3), June 2017, pp 463–78.

Text © 2024 John Wiley & Sons Ltd. Images: pp 64–5 Jelle Feringa, Joep Wijnen © Terrestrial 2023; p 66 Jelle Feringa and Diederik Veenendaal © Terrestrial and Summum Engineering 2022; p 67 Jelle Feringa © Terrestrial 2023; p 69 © Lehm Ton Erde Baukunst GmbH Emmanuel Dorsaz

Hybrid Earth-Timber Floor Slabs

Tobias Bonwetsch and Tobias Huber

Scaling Circular, Low-Carbon Construction Through Automation

Whilst reusable and sustainable, the age-old construction technique of rammed earth is also labour intensive in its compacting and tamping, and therefore expensive. The contemporary building industry consequently shies away from it. **Tobias Bonwetsch**, co-founder of Zurich-based sustainable construction start-up Rematter, and **Tobias Huber** of ZPF Ingenieure in Basel, describe their research into combining earth with locally sourced renewable timber and an automated manufacturing process to produce composite floor slabs that use the strengths and advantages of each material.

Earth is one of the oldest building materials, and rammed-earth constructions can be found all over the world, including historical earthen examples across Europe.[1] Rammed-earth structures have minimal levels of grey energy, possess unique hygroscopic properties and are 100 per cent recyclable,[2] but despite these obvious advantages, they have only a minor role in today's construction industry. This is largely due to their limited load-bearing capacity, which means they are only suitable for small buildings. In addition, the process of tamping and compacting is extremely labour-intensive, meaning that the costs of rammed-earth structures cannot compete with those of alternative construction systems.

Hybrid sustainable earth and timber floor-slab construction such as that used for the House of Research, Technology, Utopia and Sustainability (HORTUS) office building in Basel (due for completion in 2025), developed by engineers ZPF Ingenieure and architects Herzog & de Meuron with their client Senn Resources, overcomes these hurdles. The two materials are combined in such a way that their individual properties achieve their full potential, so the floor slab system can be applied to buildings of different functional typologies and various dimensions. In addition, an industrialised premanufacturing process allows highly efficient production at a competitive cost when compared to alternative floor slab solutions.

Herzog & de Meuron,
House of Research, Technology, Utopia and
Sustainability (HORTUS),
Allschwil, Basel, Switzerland,
due for completion in 2025
Sectional perspective. The rammed-earth infills of the floor slab are intentionally kept visible. This allows the earth's positive climate-control properties to be fully utilised, regulating humidity and acting as a thermal storage.

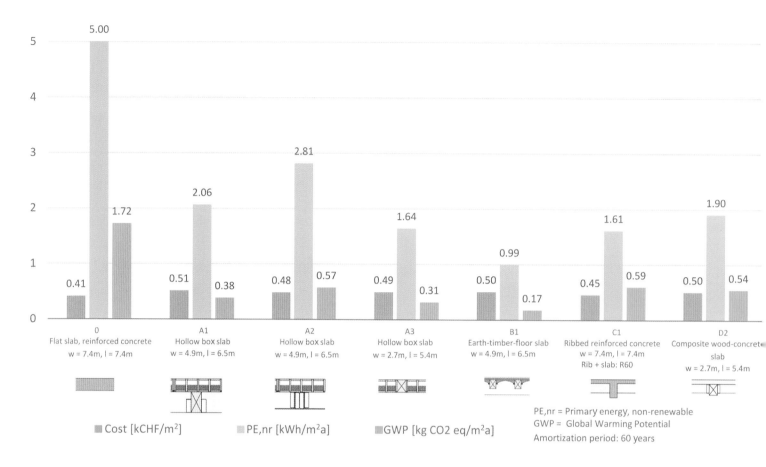

ZPF Ingenieure,
Comparison of different floor slab systems, 2020
In terms of embedded CO_2 and grey energy, the earth-timber floor slab is best in class. However, the graph also shows that intelligent material usage yields efficient structures. Still, the standard reinforced concrete slab has the most widespread use, which shows that the industry is mainly cost driven.

Floor Slabs Affect Sustainability

Today, the development of sustainable buildings and structures is not only an obligation but also a great opportunity. This is especially true for structural engineers. The elements that fall within their responsibility generally account for over 60 per cent of a building's carbon footprint.[3] The share of emissions from floor slabs in the entire load-bearing structure is just under 40 per cent.[4] This makes it clear that it is worth focusing on these components, as floor slabs offer a huge lever for improving the overall environmental footprint of a building. Obviously, the market-dominating reinforced concrete flat slab generates huge CO_2 emissions, consumes a disproportionate amount of grey energy and is difficult to dismantle. These negative aspects stem from a very inefficient use of material. However, the popularity of the concrete slab is based on simplicity in its application and relatively low construction costs. This well-known fact is highlighted in a graph showing a detailed comparison between different floor slab systems based on grey energy usage, global warming potential and their production costs. Clearly, structurally sensible component cross-sections lead to efficient load-bearing

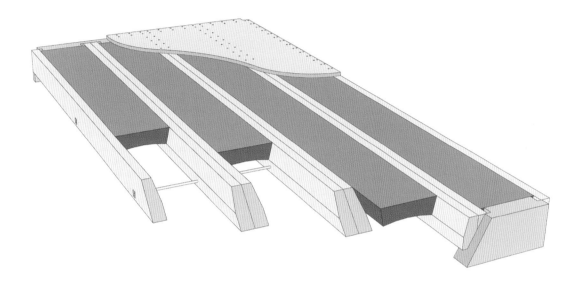

ZPF Ingenieure,
Schema of earth-timber floor slab,
2020
The system makes use of widely available, locally sourced, low-carbon materials – earth and timber. Each material is used according to its properties. The timber assumes load-bearing functions, while the earth introduces mass into the system and is only self-supporting.

structures. Moreover, when it comes to grey energy and global warming potential, the earth timber solution excels with 80 per cent less CO_2 emissions and 60 per cent less grey energy usage than a comparable flat concrete slab.

The Earth-Timber Model

Earth-timber floor slabs consist of solid timber beams with rammed-earth infills. Stiffening is achieved via a three-layer panel. All connections are dry joined or screwed, which allows for an easy disassembly and reuse of all components. No stabilisers are used for the earth, and the beams are glue-free. Combining these aspects ensures an excellent ecological balance.

The rammed earth is only self-supporting and performs as a vault. It thus follows the basic principle mentioned above of using materials according to their properties and, based on this paradigm, constructing efficient and sustainable components. Equivalent to an arched bridge made of masonry, the earth transfers the loads via pure compressive forces. Tensile or shear forces that are unfavourable for the material are completely avoided.

In order to conceal the transitions between individual floor slab elements, the timber beams are doubled with single beams at the edges. To prevent lateral deflection of the slimmer edge beams, especially during construction, a tension member is arranged at the upper and lower third point of the system. This short-circuits the horizontal shear from the arch effect within an element. Contrary to the assumption that wood would be the first choice for sustainability reasons, a steel tension member actually offers lower embodied emissions. The reason for this lies not only in the significantly higher tensile strength of the steel and the associated, much smaller cross-section, but also in the connection detail. The fasteners for a timber solution cannot be placed too near to each other or to the edge, and this requirement of minimum spacing leads to a disproportionately large cross-section of the timber member. Ultimately, this results in a worse life-cycle assessment than the minimalist steel rod.

> The construction industry needs to offer sustainable and scalable solutions to meet the climate targets and become climate-neutral by 2050

Earth-timber floor slabs meet all the requirements for multi-residential and office buildings. Firstly, the mass of the earth compensates for a common disadvantage of timber-only structures. Wood is a very efficient building material with a high load-bearing capacity in relation to its weight. However, as a consequence, this lack of dead weight means that the required sound insulation values can often only be achieved through additional measures, such as weight-bearing fills that are applied to the ceiling or as additional layers on top, which results in an overall increase of the ceiling thickness.

Further, the earth is deliberately arranged on the underside of the floor slab, where it can act as thermal mass that contributes to a passive system of night-time cooling. Simulations performed by Transsolar KlimaEngineering on an exemplar project demonstrate that the temperature-regulating properties of the floor slab regarding heat storage are equivalent to standard timber-concrete composite floors and outperform timber-only structures. Also, the thermal mass of the earth is part of the ceiling, which typically generates more applicable surface area than walls or floors that might be concealed during a building's utilisation phase and, therefore, can achieve a significantly greater degree of efficiency. In addition, the excellent moisture-regulating properties of rammed earth provide a healthy room climate.

Lastly, the earth shields relevant parts of the timber structure from direct fire exposure, which allows it to optimise the timber beams for load rather than burn-off to meet fire resistance requirements. Fire resistance is rated according to the REI scale, signifying *résistance* (resilience, or load-bearing capacity), *étanchéité* (impenetrability) and *isolation* (insulation). Currently, a targeted fire resistance of REI 60 has been verified through a large-scale fire test. This corresponds to the floor slab withstanding 60 minutes of standard fire exposure as quantified by the International Organization for Standardization (ISO), under payload. In addition, it guarantees a minimum of 60 minutes of room closure, with no smoke or flame penetration, and that the temperature rise on the side facing away from the fire will not exceed 140°C (284°F). ETH Zurich spinoff IGNIS, which has extensive experience in fire testing and performance-based verification of timber buildings, defined the test procedures and the classification of the fire resistance. These tests were essential in order to apply the earth-timber floor slab to office or multi-residential buildings. In Switzerland, combined with a sprinkler system concept, the floor slab system can be used in high-rises up to 100 metres (330 feet) tall.

Industrial Prefabrication

The construction industry needs to offer sustainable and scalable solutions to meet the climate targets and become climate-neutral by 2050.[5] These solutions also need to be economically competitive to achieve a leverage effect. Individual trailblazing projects inspire experts and laypeople alike, encourage imitation and raise awareness of the problem among the general public. As one-off projects, however, they cannot achieve a noticeably positive effect on greenhouse gas emissions.

Based on the initial development of the earth-timber floor slab, Rematter has made it its mission to spread sustainable construction and lead the earth-timber floor slab to widespread application. Specifically, Rematter is working on further developing and optimising the floor slab system and adapting it for other application areas. In addition to school and office buildings, there is a great potential for earth-timber floor slabs in residential construction in particular. The conventional spans in the latter allow for particularly efficient use of materials. Also, an EU-wide general approval of the system is in the works, which will ease application in projects outside of Switzerland. In parallel, a production line is being set up to enable industrial prefabrication of the slabs. The first ones for a multi-residential home are now being manufactured.

Rematter,
Earth-timber floor slab mock-up,
2023
opposite: Detail seen from below, showing the rammed earth in direct contact with the air within the room to maximise the benefit from the material's climate-control characteristics.

Rematter,
Robotic fabrication prototype,
ETH Zurich,
2021
above: An automated manufacturing process allows economic production of the ceiling modules.

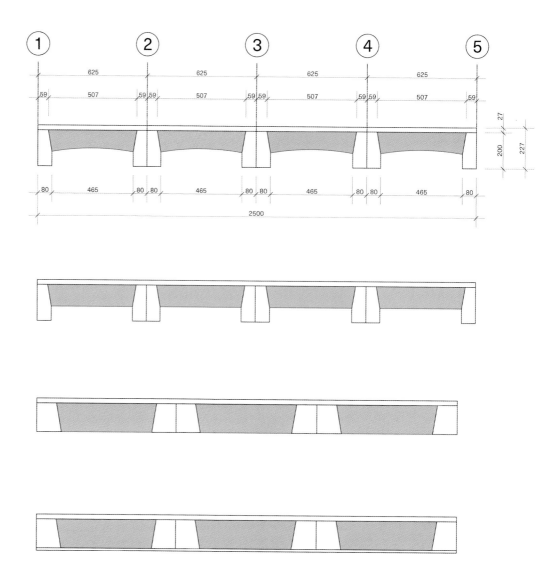

> The aim is to offer a genuine alternative to traditional and established construction systems that are the lowest in carbon emissions and are fully circular

Rematter,
Earth filling design variants,
2022
above and opposite left: Exemplar dimensions and variants of earth filling illustrate the flexibility of the system to adapt to project-specific dimensional, aesthetic and functional requirements.

Rematter,
Premanufacturing of floor slabs,
2023
opposite right: Earth-timber floor slab elements for a multi-residential housing project are produced off site. After the process of compacting the earth, the elements are ready for transport and can be installed on site.

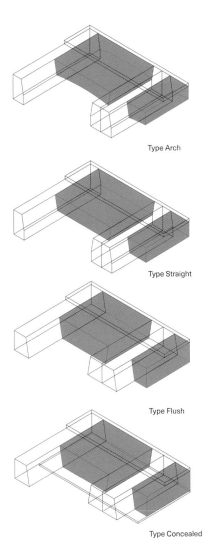

Type Arch

Type Straight

Type Flush

Type Concealed

The production steps entail the assembly of the timber frame consisting of the solid timber beams and the three-layer board. The frame acts as a 'lost formwork' for compacting the earth infills. The automated production allows the rammed earth to be processed very efficiently. This overcomes one of the biggest obstacles in today's use of this millennia-old construction method, namely the high production costs resulting from the very time-consuming manual manufacturing process. While the prototype built for the fire test was still manufactured using the traditional process with a conventional hand-operated tamper, the robotic compacting process has already yielded significant labour and time savings. As soon as the earth is compacted, the elements are ready for installation on site. No drying time is necessary. For assembly on site, the prefabricated elements are rotated 180 degrees and can then be fitted with conventional flooring layers.

Industrial production enables an economy of scale. This is key because, ultimately, the floor slab system's positive ecological and indoor climate properties are not enough to prevail in the construction industry. What primarily determines whether a building product can successfully penetrate the market, and thereby have a real impact, is its price. The aim is to offer a genuine alternative to traditional and established construction systems that are the lowest in carbon emissions and are fully circular. ⌂

Notes
1. Marlène Witry and Hubert Guillaud, 'The Distribution of Rammed Earth Construction as a Tradition in Europe', in Roger Boltshauser, Cyril Veillon and Nadia Maillard (eds), *Pisé – Rammed Earth: Tradition and Potential*, Triest Verlag (Zurich), 2019, pp 14–17.
2. Dominique Gauzin-Muller, 'The Benefits and Limitations of Earth Architecture', in Jean Dethier (ed), *The Art of Earth Architecture: Past, Present, Future*, Thames & Hudson (London), 2020, pp 24–25.
3. SIA Zurich, *SIA 2032 / 2010: Graue Energie – Ökobilanzierung für die Erstellung von Gebäuden*, SIA (Zurich), 2010, p 27.
4. Stefanie Weidner *et al*, 'Graue Emissionen im Bauwesen – Bestandsaufnahme und Optimierungsstrategien', in *Beton- und Stahlbau* 116 (12), December 2021, pp 969–77.
5. United Nations Environment Programme, '2021 Global Status Report for Buildings and Construction: Towards a Zero-emissions, Efficient and Resilient Buildings and Construction Sector', UN Environment Programme (Nairobi), 2021: https://globalabc.org/sites/default/files/2021-10/GABC_Buildings-GSR-2021_BOOK.pdf.

Text © 2024 John Wiley & Sons Ltd. Images: p 70 © Herzog & de Meuron, Hortus, Allschwil, 2020–2025; pp 72–3 © ZPF Ingenieure; pp 74, 76–7 © Rematter; p 75 © Ephraim Bieri

Look Down

Solving the Building Industry's Wasteful

Sasha Cisar

Not Up!

Conundrum with Digital and Earthen Construction

Felix Hilgert / LEHMAG,
EFH Preno housing project,
Hebertswil, Switzerland,
2022

opposite: Felix Hilgert's company LEHMAG seeks to promote the use of excavated earthen materials in construction. Excavated clay was used for prefabricated rammed-earth walls in the Hebertswil project.

above: The prefabricated rammed-earth walls are put in place at the site.

Is earth a bono fide alternative to concrete? **Sasha Cisar** believes in some cases it is. A trained architect who leads sustainability research at radicant, Switzerland's first digital sustainability bank, he shows us some recently constructed examples. The synthesis of digital technologies with these types of materiality and integrated construction techniques, and their full adoption in the building industry, could reap huge rewards and contribute to climate action.

The building and construction sector finds itself at a crossroads. The UN projects that 68 per cent of the world's population will be living in urban areas by 2050.[1] We are witnessing rapid urbanisation[2] that in part might provide much-needed housing and infrastructure. The construction sector also amounts to 5.5 per cent of gross domestic product (GDP) in the EU and approximately 13 per cent of global GDP.[3] The overall value of real-estate assets surpasses the global GDP by 2.8 times.[4]

Despite the benefits it offers, the sector significantly contributes to environmental degradation by driving climate change and biodiversity loss. According to the International Energy Agency (IEA), in 2022 the share of energy- and process-related emissions of the building and construction sector amounted to 37 per cent of global CO_2 emissions.[5] In Switzerland alone, construction accounts for 30 per cent of CO_2 emissions, 50 per cent of the raw material demand and over 80 per cent of waste produced.[6]

The deadlines for achieving the UN Sustainable Development Goals (SDGs) by 2030 and the Paris Agreement's goal of achieving net-zero CO_2 emissions by 2050 are very fast approaching. The construction sector must pivot towards more sustainable practices.

How can we change the construction industry to eliminate the wasteful material flow of earthen materials that are currently excavated and transported to landfills like gravel pits, only to be used in concrete production and returned to construction sites? The solution lies not in the skies but beneath our feet, with a blend of digital innovation and earthen construction technologies that promise to revolutionise the industry.

The Urgent Need for Change

Cement production has tripled since 1995, plateauing at around 4.4 billion tons per year after 2013.[7] Efforts to curb concrete and virgin material use persist, yet high demand strains natural resources and contributes to CO_2 emissions from resource extraction, processing and transport. Notably, construction materials account for 10 per cent of global CO_2 emissions.[8]

Construction and demolition waste amounts to a third of waste generated in the EU, and much of it lands on landfills.[9] In Switzerland, construction and demolition is by far the largest contributor of waste: excavated and quarried materials amount to two-thirds of all waste, or 57 million tonnes, and a fifth or 17 million tonnes comes from demolition waste.[10] Approximately 70 per cent of deconstruction materials are recycled.[11]

Felix Hilgert / LEHMAG,
EFH Preno housing project,
Hebertswil, Switzerland,
2022
left: Interior view of one of the prefabricated rammed-earth walls.

ERNE Holzbau and Burkard Meyer Architekten,
Extension of ERNE office building,
Stein, Switzerland,
2023
right: ERNE temporarily built a robotic prefabrication system for rammed-earth walls, shown here from inside the formwork.

A solution may well lie in one of humanity's oldest building materials: earth. Earthen construction is an ancient technique that offers a low-carbon alternative to conventional building materials like concrete and steel

ERNE Holzbau and Burkard Meyer Architekten, Extension of ERNE office building, Stein, Switzerland, 2023
View of the atrium of the ERNE office extension showcasing different types of timber and earthen construction.

Marrying Tradition with Innovation
The 'Déclaration de Chaillot'[12] that was adopted in February 2024 by representatives of 70 countries at the first ever Buildings and Climate Global Forum offers a chance to rethink construction materials and methods. It was organised by the French Government and the UN Environment Programme, bringing together actors across the building sector from governments, engineering, construction and real estate.

The declaration recognises the construction sector's massive environmental footprint, as well as the widespread disparities between building energy and climate performance and the required paths to net zero. It outlines the necessary transition towards sustainability. A key objective is the prioritisation of building materials which are on-site assets, bio- or geo-sourced and enhance the carbon balance through storage and absorption.

The Benefits of Building with Earth
A solution may well lie in one of humanity's oldest building materials: earth. Earthen construction is an ancient technique that offers a low-carbon alternative to conventional building materials like concrete and steel. When combined with modern technology, such as digital fabrication and prefabrication, it has the potential to transform the industry into a model of sustainability.

Earth can be sourced locally, thus supporting local economies and minimising emissions resulting from transportation. A substitute for conventional materials like concrete or bricks, it offers a positive impact by reducing the carbon footprint associated with building materials.

While omnipresent at construction sites, earth is, however, part of a wasteful material flow that would need to be broken: 25 per cent of earth is excavated and transported to landfills, namely gravel pits, where it is used to fill the pits to cover for the extracted gravel, which is in turn used to produce mainly concrete and thus returned to the construction site.[13]

Earth is biodegradable and non-toxic, contributing to healthier indoor air quality and reducing the overall environmental impact of buildings. Moreover, earth has excellent thermal mass properties, providing natural insulation that can reduce energy consumption for heating and cooling.[14]

Buildings in real estate have the potential to become stranded assets due to their whole-life carbon footprint, reaching obsolescence for not meeting regulatory efficiency standards, decarbonisation pathways or market expectations.[15] As a result, the value of real estate can be adversely impacted or financing impaired.[16]

In summary, earthen materials can positively impact indoor air quality and occupant health, reduce carbon emissions and thus future-proof buildings – with ancient technology.

ERNE Holzbau and Burkard
Meyer Architekten,
Extension of ERNE office building,
Stein, Switzerland,
2023

right: Juxtaposition of exposed hybrid timber construction against a rammed-earth wall, creating an interaction of the two different technologies and materials.

below: Solid walls of robotically prefabricated rammed-earth walls against the façade of timber hybrid construction.

Digital Fabrication and Prefabrication: Construction's Future

Although digital technologies have been used in construction since 1952, they have a low adoption rate.[17] Various technology types are involved, including 3D printing or robotic manufacturing, and their adoption barriers vary from economic, organisational or personnel-related to matters of policy. The potential lies not only in a single technology but within a system where it can leverage benefits, also considering significant upfront investment requirements.[18] Integrating digital technologies in construction processes and combining different technologies could be a game-changer.

Digital fabrication, including 3D printing, allows precision in the use of materials, reducing waste and optimising design for sustainability and productivity.[19] Further reduction of environmental impact and increased efficiency could be achieved through prefabrication, where the building components are assembled and transported to the construction site. Current research and projects provide a roadmap for bringing digital earth construction closer to industrial scale.[20] While there are still research gaps to close, digital earth construction could contribute towards a circular economy.[21]

Various companies are driving digital earth construction. In 2023, ERNE Holzbau and Burkard Meyer Architekten realised an office building in Stein, Switzerland, that showcases digital and prefabricated hybrid timber construction, using robotically prefabricated earthen walls to give additional structural support and provide thermal mass.[22] To prefabricate the rammed-earth walls, ERNE set up a temporary production line within its manufacturing plant.

ERNE was consulted by the civil engineer Felix Hilgert and his company LEHMAG, which seeks to make the use of excavated earthen materials in construction more mainstream. He was also involved in a 2022 housing project in Hebertswil, Switzerland, that was built in a timber hybrid construction with prefabricated rammed-earth walls using local excavation material.

ERNE Holzbau and Burkard Meyer Architekten,
Extension of ERNE office building,
Stein, Switzerland,
2023
Closed rooms were created with rammed-earth walls which are slightly detached from the timber construction.

Two start-ups in Zurich are actively promoting and advancing modern applications of earthen materials. One is rematter, which offers floor slabs as a hybrid timber construction with earth infills. The building component is designed for disassembly and to be reused. The other is Oxara, which is actively advancing the revolution of chemical admixtures to repurpose excavation materials to be used as a fluid concrete-like castable material to substitute concrete. Their technology has been implemented for several projects, including a retrofitting project in Zurich (2023) and a multi-family house in the Swiss municipality of Jona (2024). Oxara has also collaborated with student projects at ETH Zurich and at the Academy of Architecture of the Università della Svizzera Italiana (USI).[23]

Beyond Digital Earth

The adoption of digital and earthen construction faces challenges, but these could be overcome by concerted efforts, as the construction sector in Switzerland and around the world is at a crossroads. By combining digital innovation and ancient construction, it is possible to lead the way in terms of achieving positive environmental and social impacts.

The building and construction sector's journey towards sustainability is both a challenge and an opportunity. By looking down, not up, and embracing the ground beneath our feet as a source of inspiration and material, the industry can transform itself.

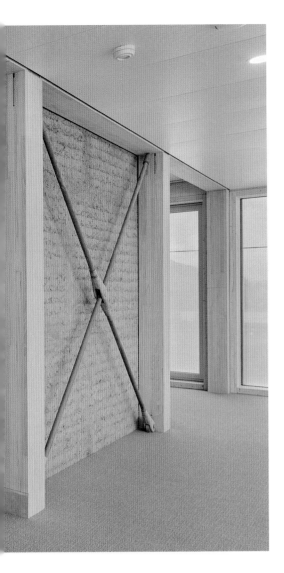

Notes

1. *World Urbanization Prospects: The 2018 Revision*, UN Department of Economic and Social Affairs (New York), 2019, p 1.
2. 'The World is in the Midst of a City-Building Boom', *The Economist*, 7 March 2024: www.economist.com/finance-and-economics/2024/03/07/the-world-is-in-the-midst-of-a-city-building-boom.
3. Filipe Barbosa et al, *Reinventing Construction Through a Productivity Revolution*, McKinsey Global Institute, 27 February 2017, p 1: www.mckinsey.com/industries/capital-projects-and-infrastructure/our-insights/reinventing-construction-through-a-productivity-revolution.
4. Paul Tostevin and Charlotte Rushton, 'Total Global Value of Real Estate Estimated at $379.7 trillion – Almost Four Times the Value of Global GDP', Savills, 25 September 2023: www.savills.com/insight-and-opinion/savills-news/352068/total-global-value-of-real-estate-estimated-at-$379.7-trillion---almost-four-times-the-value-of-global-gdp.
5. *Global Status Report for Buildings and Construction*, United Nations Environment Programme / Global Alliance for Buildings and Construction (Nairobi), 2024, p 17: www.unep.org/resources/report/global-status-report-buildings-and-construction.
6. Swiss Federal Office for the Environment, 'Raw Materials, Waste and the Circular Economy: In Brief', updated 30 November 2018: www.bafu.admin.ch/bafu/en/home/topics/waste/in-brief.html.
7. 'Production Volume of Cement Worldwide from 1995 to 2023', Statista, 2023: www.statista.com/statistics/1087115/global-cement-production-volume/#:~:text=The%20total%20volume%20of%20cement,industry%20has%20grown%20since%20then.
8. Swiss Federal Office for the Environment, *op cit*.
9. 'Construction and Demolition Waste', EU Commission, 2018: https://environment.ec.europa.eu/topics/waste-and-recycling/construction-and-demolition-waste_en#:~:text=Construction%20and%20demolition%20waste%20(CDW,%2C%20glass%2C%20metals%20and%20plastic.
10. Swiss Federal Office for the Environment, *op cit*.
11. *Ibid*.
12. 'Declaration de Chaillot – Forum Mondial Bâtimoent et climat', UN Environment Programme, 8 March 2024: www.unep.org/news-and-stories/press-release/buildings-and-climate-global-forum-declaration-de-chaillot.
13. Stefan Rubli, 'KAR-Modell – Modellierung der Kies-, Rückbau- und Aushubmaterialflüsse: Nachführung Bezugsjahr 2022', Kanton Zürich (Zurich), 2023.
14. Lola Ben-Alon and Alexandra R Rempel, 'Thermal Comfort and Passive Survivability in Earthen Buildings', *Building and Environment* 238 (4), April 2023, 110339, and Pragya Gupta et al, 'Evaluation of Rammed Earth Assemblies as Thermal Mass through Whole-Building Simulation', 2020 Building Performance Analysis Conference and SimBuild co-organised by ASHRAE and IBPSA-USA, Chicago, 2022.
15. Sally Anne Stone, 'Stranded Real Estate – A Real Possibility', Matheson, 12 June 2023: www.matheson.com/insights/detail/stranded-real-estate-a-real-possibility.
16. Michael Armstrong and Rebekah Mulryan, 'The Future of Commercial Real Estate is Written in Green: Debt Financing the Transition', Ernst & Young (Dublin), 2022.
17. Tan Tan, Ming Shan Ng and Daniel M Hall, 'Demystifying Barriers to Digital Fabrication in Architecture', Engineering Project Organization Conference, Berlin, 2023.
18. Katy Bartlett et al, 'Rise of the Platform Era: The Next Chapter in Construction Technology', McKinsey & Company, 30 October 2020: www.mckinsey.com/industries/private-equity-and-principal-investors/our-insights/rise-of-the-platform-era-the-next-chapter-in-construction-technology.
19. Isolda Agustí-Juan, Andrei Jipa and Guillaume Habert, 'Environmental Assessment of Multi-Functional Building Elements Constructed With Digital Fabrication Techniques', *The International Journal of Life Cycle Assessment* 24, June 2019, pp 1,027–39, and Borja García de Soto et al, 'Productivity of Digital Fabrication in Construction: Cost and Time Analysis of a Robotically Built Wall', *Automation in Construction* 92, August 2018, pp 297–311.
20. Mohamed Gomaa et al, 'Digital Manufacturing for Earth Construction: A Critical Review', *Journal of Cleaner Production* 338, January 2022, 130630.
21. Marcel Schweiker et al, 'Ten Questions Concerning the Potential of Digital Production and New Technologies for Contemporary Earthen Constructions', *Building and Environment* 206, 2021, 108240.
22. Patrick Suter and Fabian Hörmann, 'Digitalization and Prefabrication', in Fabian Hörmann (ed), *The Real Deal: Post-Fossil Construction for Game Changers*, Ruby Press (Berlin), 2023, pp 146–51.
23. 'USI: Earth-to-Earth / Ultra-Thin Formwork For In-Situ Casting': www.arc.usi.ch/en/feeds/12764.

The combination of digital and earthen construction offers a path forward that aligns with sustainable development goals, aiming for net-zero CO_2 emissions, minimising waste, and creating sustainable, healthy living spaces. The future of construction is already here. ⌂

By combining digital innovation and ancient construction, it is possible to lead the way in terms of achieving positive environmental and social impacts

Text © 2024 John Wiley & Sons Ltd. Images: pp 78–9, 81(l) © Felix Hilgert; p 81(r), 82–6 © ERNE AG Holzbau, Stein (CH) | Bernhard Strauss, Freiburg (DE)

Corentin Fivet, Stefana Parascho, Maxence Grangeot and Malena Bastien-Masse

Disused

Concrete, Digital Acupuncture and Reuse

Maxence Grangeot,
Stefana Parascho and
Corentin Fivet,
Digital Upcycling of
Concrete Rubble,
Ecole Polytechnique
Fédérale de Lausanne
(EPFL),
Switzerland,
2023

Upcycling concrete waste
into structures by using
digital fabrication can provide
both circular solutions for
construction and new tectonics
for architects.

Maxence Grangeot,
In-situ survey of recycling centres,
Ecole Polytechnique Fédérale de
Lausanne (EPFL),
Switzerland,
2023

Most demolished concrete is gathered in recycling centres. Large pieces of rubble are crushed further to fit the crusher's intake. If not landfilled, the crushed concrete replaces portions of natural aggregates in new concrete mixes. Instead, the debris could be reused as-is in new architecture or structural applications.

More digitalisation, more sustainability, more circularity, more well-being for all: the aspirations fuelling contemporary societal development are well known, and research projects are multiplying in an attempt to satisfy them. However, while a lot *can* be done, what *needs* to be done is still open to debate. Recent research undertaken at the Ecole Polytechnique Fédérale de Lausanne (EPFL) on new load-bearing applications for concrete waste showcases how targeted digital interventions could lead to immediate and exceptional environmental benefits. New dynamics between professionals and their tools are needed. In reaction to usual technology-driven research, such mission-oriented research calls for new pragmatic blends of low and high technology within established mainstream routines.

Waste-based Circular Construction

Concrete is a ubiquitous construction material with unrivalled qualities. It is cherished by designers, constructors and land developers. Given the immense detrimental effects of its

Crushing concrete liberated from demolished buildings to provide recycled aggregate is argued to be a sustainable way of reducing material extraction. Yet, in these high-energy processes, large amounts of new cement is often used, negating this as a low-carbon activity. Professor of Architecture and Structural Design at the Ecole Polytechnique Fédérale de Lausanne (EPFL) **Corentin Fivet** and his co-authors highlight the advantages of utilising second-hand uncrushed concrete while augmenting traditional equipment to remarkable tactile and optimum effect.

production on global warming and natural ecosystems,[1] it can even be argued that concrete is not cherished enough. Indeed, driven by functional obsolescence or real-estate pressures, premature demolition of concrete structures generally arises far too often, and at a time when they are still performing well and present only inconsequential degradation, if any.

The industry is currently disregarding the millions of tonnes of this continuously flowing high-quality waste material. Currently, only a fraction of all rubble from demolition is crushed into smaller pieces that are then used as aggregate in newly mixed 'recycled' concrete. Although this recycling operation allows for the partial replacement of raw natural aggregates, it still requires equivalent amounts of cement. It has, therefore, nearly zero impact on reducing global warming potential (GWP) and should be delayed as much as possible.

Instead, avoiding the crushing of obsolete concrete offers the opportunity to reclaim its pre-existing mechanical and functional features in new applications. This other circularity process is known as 'component reuse'. It aims to rechannel waste flows from demolition sites directly to new construction sites without introducing large, irreversible transformative operations, to substitute the production of new components without reducing quality expectations.

When it comes to disused concrete structures, two deconstruction methods stand out. The most common and cheapest involves conventional hydraulic shears, offering reuse applications through masonry-like assemblies. The other, saw-cutting, though pricier, preserves original geometric and structural features, allowing substantial raw material savings in new construction, which in return generally balances the overall financial budget. This practice is not new: a recent historical review identified 77 precedents, including the dismantling of multistorey buildings and the reuse of their panels as load-bearing walls and slabs in new residential buildings.[2]

... computational algorithms play a crucial role in quickly finding optimal matches between an inventory of reusable components and the expected layout (as provided by the Phoenix3D plugin for Rhino3D)

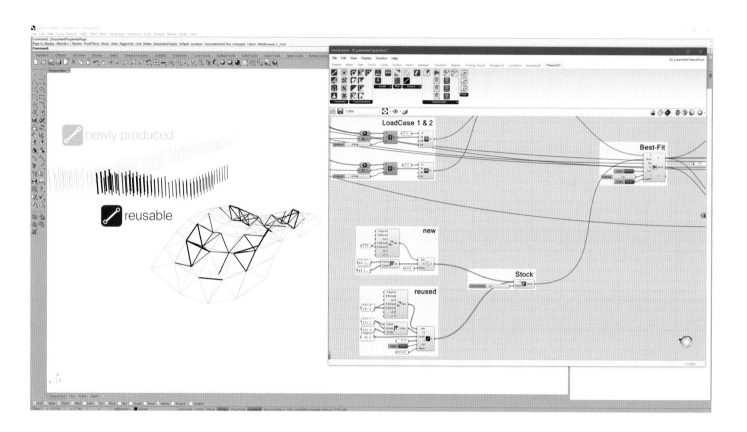

Beyond the Hype and the Niche

Systematic urban mining of disused concrete for reuse provides significant environmental benefits while offering capabilities similar to prefabricated concrete. The approach would reduce the GWP of new structures by up to 90 per cent,[3] avoid extraction activities, and drastically cut waste volumes beyond what other reused building materials achieve. These benefits come in addition to the many positive side-effects of component reuse activities in general. They foster the creation and maintenance of local and non-externalisable jobs, and they lower the dependence on imports and international trade fluctuations. Moreover, they encourage a fresh perspective on built heritage in which components evoke a longer story than the building they are part of, and in which patina and traces of past use are given more value than the uniformity of pristine, serialised products.

Despite all of its positive impacts, the reuse of load-bearing components, especially concrete, can only be a palliative strategy. Decommissioning and building anew should be a last resort after every effort has been made to keep existing buildings in place and adapt them in situ. Nevertheless, in the face of the climate emergency, embracing the reuse of disused concrete components is a necessary strategy. This holds true as long as demolition and construction rates do not decrease, cement plants are not fully decarbonised, and carbon-free building materials cannot replace all applications of concrete.

Spearheading the circular economy and gaining traction with governments and lobbies at all levels, component reuse activities are burgeoning globally. However, they predominantly involve non-load-bearing elements, such as those from the building envelope, finishing or equipment. Although this represents a positive development, the impact on climate crisis alleviation remains minimal. The amount of potentially saved CO_2 will remain negligible compared to concrete production, and acclaimed practices of component reuse in architecture may remain confined to niche markets. The same is probably true about many recent academic projects on circular construction, involuntarily often contributing more to greenwashing than to tangible techno-societal improvement. In this context, the challenge is less about identifying new circular gimmicks and more about finding ways to incrementally strengthen mainstream professional dynamics towards more circular and sustainable routines.

Jonas Warmuth, Jan Brütting and Corentin Fivet,
Phoenix3D for Rhino3D and Grasshopper,
Ecole Polytechnique Fédérale de Lausanne (EPFL),
Switzerland,
2021

opposite: Developed at EPFL, Phoenix3D is a plugin for Rhino3D and Grasshopper that allows users to find the best matches between an inventory of steel or timber bar elements and a desired structural layout. It optimises three questions simultaneously: what elements are worth taking from the inventory, where they should be placed in the final structure, and what shape should the final structure have? While ensuring constraints of strength and static equilibrium, it can work with different objective functions, for example minimum GWP, minimum excess cut, and minimum number of cuts.

Nicole Widmer, Malena Bastien-Masse and Corentin Fivet,
Structural Design for Reuse of Sawn Cast-In-Place
Reinforced Concrete Components,
Ecole Polytechnique Fédérale de Lausanne (EPFL),
Switzerland,
2022

below: Two disused reinforced-concrete office buildings were selected as donor structures of a new office building whose floor plan is given. Optimisation algorithms deal with finding the best cutting layout for the donor structures and the best rearrangement of the saw-cut slabs to satisfy the spans of the receiving building. Variations of bending capacities (colours in the middle figures) are taken into account, and the addition of ultra-high-performance fibre-reinforced concrete (green and blue colours in the last figure) is considered to strengthen the reused slabs where needed. This component-reuse strategy decreases GWP by 77 per cent compared to conventional new construction.

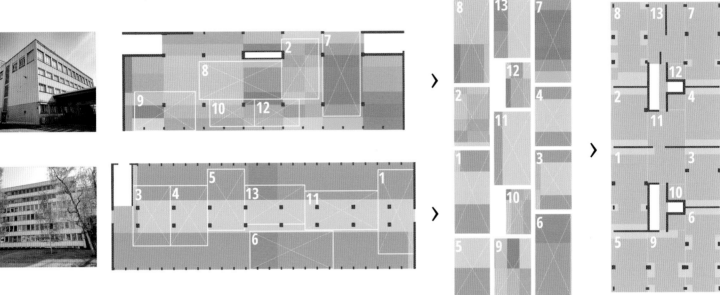

Optimum Material Reuse

Incorporating reclaimed products in a building project poses new challenges for architects and structural engineers. Designers must deal with additional uncertainty related to material availability (what quantities?), material properties (what geometry? what quality? what strength?) and variability between objects of the same stock (what tolerance? what diversity?). While resolving these uncertainties may happen very late in the design process, early design decisions can already anticipate that uncertainties will arise later in the process and seize the opportunity to mitigate their impact on the resulting construction logistics. This is where computational algorithms play a crucial role in quickly finding optimal matches between an inventory of reusable components and the expected layout (as provided by the Phoenix3D plugin for Rhino3D),[4] assessing how design choices are resilient to inventory changes, and providing fast estimates of component availability in a given donor building stock.

Optimising the assignment of reused steel or timber members in new truss structures[5] or building frames[6] can reduce the carbon footprint by up to 60 per cent. Further savings occur when replacing oversized reused members with newly manufactured ones.[7] Utilising each component to its full potential, and thus avoiding cascading downgrades wherever possible, is essential, since it allows for an efficient use of resources and minimises the need for new production. Along the same lines, allowing new greenhouse gas emissions to strengthen a reclaimed object may not be detrimental to the overall carbon footprint of the project. For instance, using ultra-high-performance fibre-reinforced concrete to strengthen reclaimed saw-cut slabs may ensure their reuse in a new building while satisfying contemporary standards of structural safety and still decreasing global warming potential by 77 per cent.[8] These early results suggest that one should not be dogmatic about circularity and that a 100 per cent reuse rate in new projects is not required to achieve significant environmental savings.

Malena Bastien-Masse, Julie Devènes
Célia Küpfer, Jan Brütting and
Corentin Fivet,
Re:Crete footbridge,
Ecole Polytechnique Fédérale de
Lausanne (EPFL),
Switzerland,
2021

above: Re:Crete is a 10-metre (32-foot) long post-tensioned footbridge, the blocks of which are reclaimed saw-cut segments of walls from a building undergoing major renovation. It is now used outdoors in Switzerland. A life-cycle assessment showed that its construction presents a GWP three to four times smaller than the construction of similar footbridges in recycled concrete or steel, and a GWP equivalent to a timber alternative.

Anticipating material variations influences the conceptual design. For instance, the 10-metre (32-foot) span Re:Crete footbridge developed at EPFL in 2021 is made of 25 concrete blocks saw-cut from the basement walls of a building undergoing major renovation.[9] The blocks exhibited variations of up to 2 cm (0.8 inches) in length, which stemmed from construction and saw-cutting tolerances. Integrating these uncertainties early in the design process by sorting the blocks allowed the minimisation of dimensional differences between adjacent ones, further helped by choosing an assembly method that allowed mortar joints to have varying widths.

Another recent application of structural concrete reuse is the student-led EPFL project rebuiLT pavilion (2023), which features a skeleton assembled from saw-cut modules consisting of two slabs connected by a mushroom column. Other projects also reclaimed concrete pieces from the same donor building,

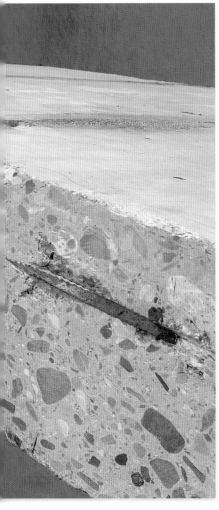

rebuiLT pavilion,
Ecole Polytechnique Fédérale
de Lausanne (EPFL),
Switzerland,
2023

below: This student-led MAKE project involved producing a small community building while employing only low-tech and reused components. Its load-bearing system is made of saw-cut assemblies of slabs and their connecting columns, extracted from a nearby building doomed to demolition. Reclaimed roof tiles and straw-bale walls complete the design.

leading to a new design task consisting of planning and optimising the deconstruction process. This is done by finding the best cutting pattern that satisfies transport constraints and element dimensions set by the various receiver buildings. The innovation here lies not in devising new fabrication techniques and then searching applications, but in first identifying relevant societal challenges – the environmental burden of demolition waste and new construction – and then exploring new design workflows that tackle them.

Digital Acupuncture

Robotic systems are generally not yet well suited for on-site day-to-day concrete construction due to reasons such as payload limits, mobility, safety, initial financial investment or required expertise. Before imposing state-of-the-art robotic devices on construction sites, much can be done by digitally augmenting conventional construction equipment (cranes, hammer drills, concrete saws) with off-the-shelf digital sensors, processors and controls. Furthermore, rather than automating entire processes, targeted digital inputs could leave room for more organic cooperation between machines and operators. They can foster readjustments arising from assembly constraints or material inaccuracies. This was demonstrated in the Zero Waste Project, a digital upcycling experiment at Princeton University in which a timber-frame structure was transformed on-the-fly by means of robotic arms.[10] Similar to acupuncture, in which specific body areas are pierced with needles for therapeutic purposes, digitalisation could be applied to only specific tasks of mainstream routines with the aim of improving sustainability or economic viability.

Another low-tech yet digital approach was recently experimented with while fabricating load-bearing walls with reused concrete rubble obtained from demolition sites.[11] This EPFL project uses a heterogeneous stock of concrete rubble pieces, which are digitised using an edge detection algorithm that is applied to pictures captured from a crane's trolley. The large and mostly flat rubble pieces are then assembled on their thin faces to construct slender walls. Particular care is given to structural robustness, optimum logistics and minimal storage by retaining adaptability in case of an unpredicted change in rubble geometry or modification of the total stock. The 2D geometries of the fragments are digitally arranged using a stacking algorithm, providing numerous solutions to choose from. The void area between large rubble pieces is minimised to reduce the amount of added material needed to provide airtightness and continuity of forces throughout the structure. Each piece of rubble is lifted from a precisely defined anchor point, and its orientation is therefore achieved thanks to gravity only, which is compatible with all existing lifting equipment. The anchor point, computed by estimating the centre of gravity of the rubble unit and its desired orientation, is precisely drilled using an off-the-shelf concrete hammer drill converted into a robotic end-effector.

The latest iteration of these concrete rubble masonry walls showcases a drastic reduction in GWP compared to manually assembled irregular stone masonry walls or recycled concrete walls. These savings are attributed to both the reuse of concrete debris and the digital augmentation of only a selected subset of the steps making up the construction process. Contrasting with all-digital academic experiments, this project illustrates that 'acupuncture digitalisation' with off-the-shelf apparatus may be what the industry actually needs at the moment. Closer collaboration between blue and white collars is required to explore the viable use cases.

Mission-oriented Research and 'Pragmatic Tech'

Mission-oriented research, as portrayed in the examples here, aims to find solutions that have a measurable positive impact on today's prominent societal problems, such as climate change or social inequalities. Instead of seeking knowledge for its own sake, the challenge is to solve complex and interdisciplinary problems in real-world contexts scientifically. Too few digital fabrication research projects fall into this category. While digital technologies hold great potential, current knowledge production often seems to be getting trapped in the belief that any high-tech development will ultimately be relevant to society. The urgency and complexity of today's global challenges do not allow for such a gamble. On the contrary, mission-oriented solutions call for a new blend of low and high technology, where industry habits and out-of-the-lab expertise are combined with a pragmatism that is yet to be explored. 🗅

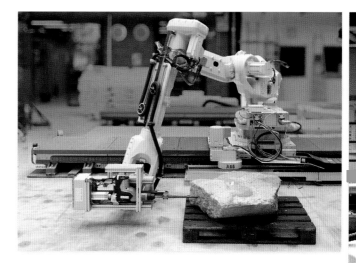

Notes
1. United Nations Environment Programme, *2022 Global Status Report for Buildings and Construction: Towards a Zero-emission, Efficient and Resilient Buildings and Construction Sector*, UNEP (Nairobi), 2022, p 42.
2. Célia Küpfer, Maléna Bastien-Masse and Corentin Fivet, 'Reuse of Concrete Components in New Construction Projects: Critical Review of 77 Circular Precedents', *Journal of Cleaner Production* 383 (3), January 2023, 135235.
3. Célia Küpfer, Numa Bertola and Corentin Fivet, 'Reuse of Cut Concrete Slabs in New Buildings for Circular Ultra-Low-Carbon Floor Designs', *Journal of Cleaner Production* 448, April 2024, 14156.
4. Jonas Warmuth, Jan Brütting and Corentin Fivet, 'Computational Tool for Stock-Constrained Design of Structures', *Proceedings of the IASS Annual Symposium 2020/21*, Guildford, August 2021.
5. Jan Brütting *et al*, 'Design of Truss Structures Through Reuse', *Structures* 18, April 2019, pp 128–137.
6. Jan Brütting *et al*, 'Optimum Design of Frame Structures From a Stock of Reclaimed Elements', *Frontiers in Built Environment* 6, May 2020.
7. Jan Brütting *et al*, 'Environmental Impact Minimisation of Reticular Structures Made of Reused and New Elements Through Life Cycle Assessment and Mixed-Integer Linear Programming', *Energy and Buildings* 215, May 2020, 109827.
8. Nicole Widmer, Maléna Bastien Masse and Corentin Fivet, 'Building Structures Made of Reused Cut Reinforced Concrete Slabs and Walls: A Case Study', in Fabio Biondini and Dan M Frangopol (eds), *Life-Cycle of Structures and Infrastructure Systems*, CRC Press (London), 2023, pp 172–9.
9. Julie Devènes *et al*, 'Re:Crete – Reuse of Concrete Blocks From Cast-in-Place Building to Arch Footbridge', *Structures* 43 (11), September 2022, pp 1854–67.
10. Edvard PG Bruun *et al*, 'ZeroWaste: Towards Computing Cooperative Robotic Sequences for the Disassembly and Reuse of Timber Frame Structures', in *Proceedings of the 42nd Annual Conference of the Association for Computer Aided Design in Architecture*, Philadelphia, Pennsylvania, 2022, pp 586–97.
11. Maxence Grangeot, Corentin Fivet and Stefana Parascho, 'From Concrete Waste to Walls: An Investigation of Reclamation and Digital Technologies for New Load-bearing Structures', *Journal of Physics Conference Series* 2600 (19), 2023, 192019.

Maxence Grangeot, Stefana Parascho and Corentin Fivet,
Digital Upcycling of Concrete Rubble,
Ecole Polytechnique Fédérale de Lausanne (EPFL),
Switzerland,
2023

opposite: An off-the-shelf concrete hammer drill is mounted at the end of a robotic arm via a shock absorber. Holes are drilled on the thin faces of rubble pieces to insert lifting anchors and thus orient the flat rubble to the desired vertical position using gravity.

above: A full-scale masonry wall made of large, flat pieces of concrete rubble from demolition using digital arrangement and fabrication. The wall is airtight, flat on both sides and withstands usual loads. The materials and process to make such a wall drastically reduce the GWP of concrete and masonry walls.

Text © 2024 John Wiley & Sons Ltd. Images: pp 88–90, 96–7 © Maxence Grangeot, EPFL; pp 92–4 © Corentin Fivet, EPFL; p 95 © Pierre-Jean Renaud – rebuiLT

Knowledge Production in

Romana Rust and Inés Ariza

Gramazio Kohler Research,
Acoustic Panel System,
Immersive Design Lab, ETH Zurich,
2021
The 3D-printed acoustic panels mounted in the Immersive Design Lab improve the acoustic quality of the space within defined frequency bands, complementing absorption panels to create a homogeneous, acoustically isotropic room for spatial audio. The developed design was patented as software, enabling reuse, and was passed on to fabrication partner Aectual, who now offer the panels as a mass-customisable product in their store.

Digital Design and Fabrication

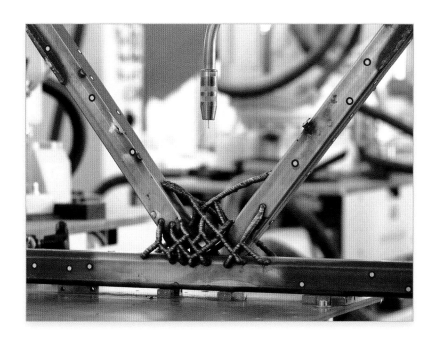

Romana Rust and Inés Ariza illustrate how digital fabrication projects developed by Gramazio Kohler Research at ETH Zurich are using innovative software pipelines and producing more collaborative software and digitally driven techniques to illicit much more dynamic and sustainably effective interfaces between machines and humans for the construction industry.
Rust is a computational architect and Head of Research and Innovation at vyzn AG, and Ariza is an architect and researcher within the NCCR Digital Fabrication consortium at Gramazio Kohler Research, and R&D Project Lead at MESH AG.

Over the last decades, the rise of the digital building culture has fundamentally changed the role of the architectural model, transforming it from a mere representation of design intent into an operative tool to build, analyse and collaborate.[1] Within digital fabrication projects, descriptive building drawings have been replaced by prescriptive rules to build, measure and adapt, leaving behind a conglomerate of software and data repositories after their realisation. But beyond being reproducible, what is the prospective value of these scripted models? What should be shared if the goal is not to replicate a project – something rare in architecture – but to tap into the expertise embedded in creating it?

While a genuine interest in finding alternative methods to minimise embodied carbon emissions has settled across architectural domains, less attention has been drawn to the underlying knowledge infrastructures that would enable such systemic changes to propagate. Today, the discourses of sustainability are strongly tied to material flows, minimisation of material footprints, and performance optimisation. However, for these to become ubiquitous, we must fundamentally rethink how we produce, share and apply disciplinary architectural knowledge. Here is where the concept of software as a repository for collective knowledge to be preserved, reused, reduced and recycled becomes increasingly critical.

Inés Ariza / Gramazio Kohler Research, Circular Engineering for Architecture and Chair of Steel and Composite Structures, Adaptive Detailing for Reclaimed Steel, ETH Zurich,
2023
opposite: Reclaimed steel parts are given a second life using a highly malleable additive joining process based on wire and arc additive manufacturing (WAAM). The position of discrete metal drops accounts for what the robot can reach, where force transfer is required, and how the metal solidification process occurs, adapting to the variable and uneven surfaces of reused profiles. These knowledge domains are integrated through a COMPAS software pipeline.

Inés Ariza / Gramazio Kohler Research, Adaptive detailing with wire and arc additive manufacturing, ETH Zurich,
2022
left: The robotic WAAM process is used here to join parts assembled in non-regular configurations. Detailing for non-standard robotic processes entails an explicit understanding of functional, material and hardware constraints.

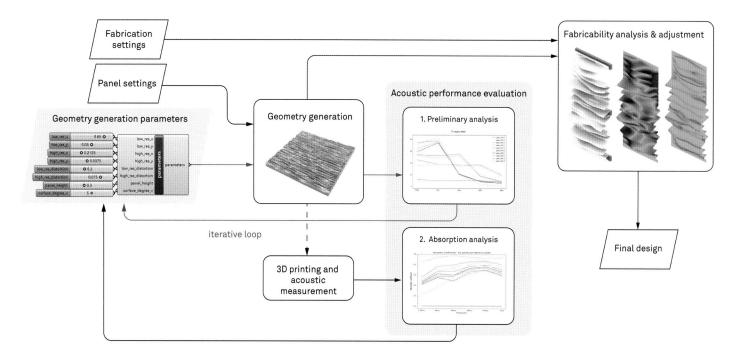

The following digital fabrication projects developed by Gramazio Kohler Research at ETH Zurich exemplify various strategies for realising this vision, showcasing how their design and production software pipelines and collaborative interfaces are paving the way to a more explicit, retrievable and accessible building culture.

Operative Pipelines and the Consolidation of Knowledge Domains

For digital fabrication in architecture, a fundamental challenge lies in merging technical know-how with disciplinary architectural expertise. This demands designers to speak a diverse array of languages – such as robot operating systems, hardware communications protocols or material specifications – leaving them no option but to navigate through other disciplines to attempt architecturally significant outcomes.

Consider the intersection of robotic manufacturing and 3D printing with detailing.[2] Specifying how parts connect through detailing is a core architectural task, and details are a common source of construction know-how. A robotically 3D-printed detail is, however, unfamiliar to architecture's knowledge base. Let us look at the knowledge that accumulates in creating such an object.

We should first observe that both the design and production pipeline and its outcome – the object – are ideated simultaneously and, inevitably, inform each other. In that sense, knowledge flows both ways from design to fabrication and vice versa: design intent factors in structural, robotic, printing and material constraints. Each planning phase integrates the creators' knowledge into a software component, taking production feasibility, structural performance and durability of the final product into account. Once a functional pipeline is established, bidirectional communication between the software that creates the object, and the machine that executes it, allows each stage of its production to be recorded. Within this feedback loop, as-built information updates the model state, storing modifications and descriptions of alterations made during production – explicit data on errors that can be used to re-create failure patterns for refining the design and production process.

The trail of data generated and measured during the manufacturing process narrates a journey from conception to realisation. This data tells us about the quality of the parts in relation to the recorded process parameters through the found tolerances or their expected durability. Yet what of this do we need to keep, if

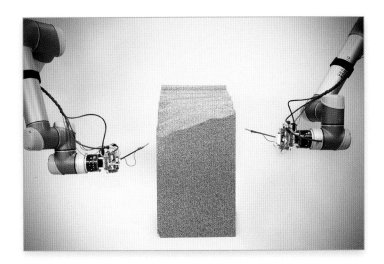

we are seeking to preserve the know-how behind the object? How would a detail magazine even illustrate it? Beyond raw data such as material coordinates or energy usage, the software manoeuvring the design and manufacturing processes is crucial for understanding their specifications and ensuring their preservation. If we want to protect the knowledge that came into this object, it is not enough to preserve the physical artefact, nor its data or digital twin. Rather, all the phases of the digital model now turned into a reusable software pipeline need to be preserved. This brings together two concepts of sustainability, both enabled by the choice of using custom design and production software: the reusability of the software components, and the possibility of extending the object's lifetime due to the availability of fully fledged material specifications.

A software pipeline in digital fabrication embodies a hybrid of knowledge domains, including the interdependencies of hardware specifications, material requirements and performance evaluations. A prime example of this integration is seen in the development of the Acoustic Panel System.[3] Here, an interdisciplinary approach brings together the acoustic performance and fabricability of diffusive panel shapes into a single computational design pipeline. This data-driven method for designing acoustics parallels the approach taken with the Spatial Wire Cutting design and production pipeline.[4] In this latter case, the cooperative robot manipulation of the wire and the material behaviour of the cut volumes explicitly inform the double-curved surface properties of the wall panels created for the Swisspearl® Summerschool.

Romana Rust / Gramazio Kohler Research,
Workflow elements and information flows of the Acoustic Panel System,
ETH Zurich,
2021
opposite: The workflow represents the integration of several knowledge domains, including a computational design system, a preliminary acoustic assessment method and the evaluation of acoustic performance through measurements and post-processing design instances to ensure fabricability. The research project has been partnered with acoustic experts RocketScience AG and Strauss Elektroakustik GmbH, as well as fabrication partner Aectual.

Romana Rust / Gramazio Kohler Research,
Experimental setup of the Spatial Wire Cutting process,
ETH Zurich,
2014
above: This special robotic wire-cutting technique creates a certain family of double-curved surfaces. They can be conceived using bespoke design and simulation software that encapsulates explicit knowledge of the material behaviour, hardware and robotic setup.

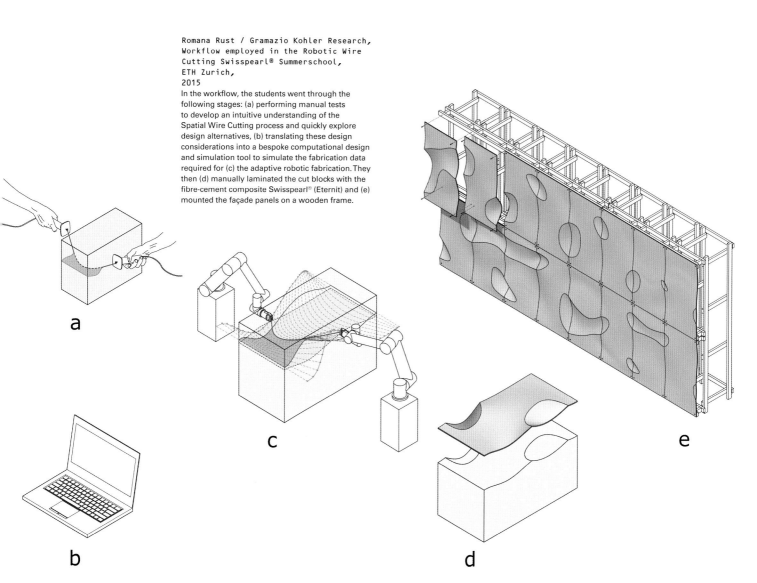

Romana Rust / Gramazio Kohler Research, Workflow employed in the Robotic Wire Cutting Swisspearl® Summerschool, ETH Zurich, 2015

In the workflow, the students went through the following stages: (a) performing manual tests to develop an intuitive understanding of the Spatial Wire Cutting process and quickly explore design alternatives, (b) translating these design considerations into a bespoke computational design and simulation tool to simulate the fabrication data required for (c) the adaptive robotic fabrication. They then (d) manually laminated the cut blocks with the fibre-cement composite Swisspearl® (Eternit) and (e) mounted the façade panels on a wooden frame.

This broad and, at the same time, specific knowledge of structural engineering, acoustics, energy simulations, material properties, hardware and software requires a dedicated framework for communication, where interfaces between domains have a special role. Emerging from an interdisciplinary research community, the COMPAS framework[5] stands as an example of a collaborative software culture. Its core library provides data structures shared across packages for robotic fabrication, additive manufacturing and structural engineering and reused in multiple research projects. These data structures serve as the backbone of domain-specific knowledge housing interdependent attributes and methods, describing object properties and possible functionalities while enabling seamless communication among software packages through industry-standard serialisation formats.

Multimodal Interfaces for Knowledge Interchange
However, not all design aspects can be embedded or extracted into software. In the end, architectural production revolves around communication between humans applying their expertise and experience, which relies on tacit knowledge and may not easily be expressed. With the digital model being the absolute project reference in today's architecture production, discussions within teams can be prone to misconceptions when the model is inaccurately perceived as the ultimate source of truth or if not all involved parties share the same design representation or context.

A potential resolution involves extending the modality of communication between on-site and off-site users to facilitate bidirectional data exchange and decision-making. This is the case of extended-reality environments that provide means for users to meet, visualise and interact with digital objects, either by creating an artificial environment (virtual reality) or by superimposing digital elements onto the real world (mixed reality). Prototypical multimodal environments such as the Extended Reality Collaboration (ERC) system[6] grant users access to any relevant conversation channel to manipulate data and evaluate design iterations dynamically. This interface, in turn, can allow design discussions to be traced to a moment in time, linked to a concrete piece of information and identified stakeholders, similarly to how code is traced by version control. In a second step, these systems can be extended to augment the construction phases so that the state of the building site can inform design decisions and new design iterations can be evaluated on site instantaneously.

This is foreign to conventional project management in architecture, where the design process starts on site but an abstraction of real conditions is quickly created. First impressions or recorded conditions are accepted as valid throughout the project. In most cases, however, these conditions are dynamic and subject to change. With multimodal collaborative environments, discrepancies between the as-built and the digital model can be easily detected, which is particularly crucial for modifications, revisions and approval phases. The open channel to the construction site and its status becomes a common source of truth. This online repository can be referred to any time to mitigate compartmentalised decision-making rooted in incomplete or inaccurate data. The continuous input from the site and, vice versa, the continuous validation of the design on site creates a knowledge-building dialogue around the consecutive and interdependent design decisions, as well as a knowledge-sharing platform collectively managed by different stakeholders at different locations with varying expertise.

Daniela Mitterberger / Gramazio Kohler Research,
Scenarios for knowledge interchange, ETH Zurich,
2022
Through an extended-reality system, incorporating both virtual- and mixed-reality devices, users can preview and annotate design options on site and receive and request site information to supervise fabrication phases when off site.

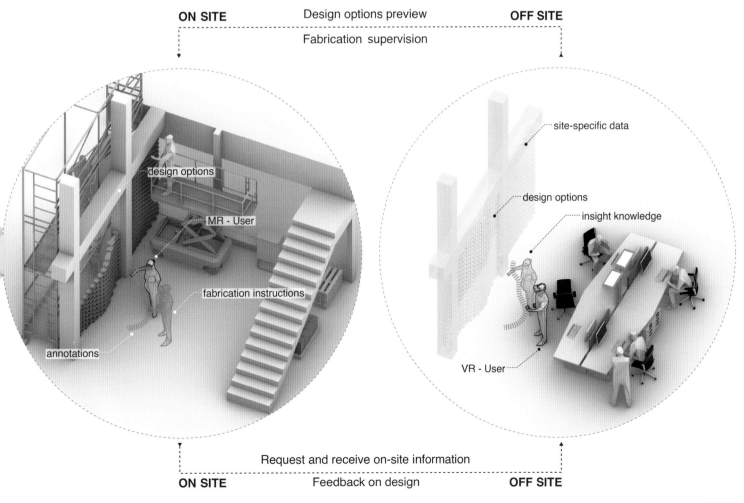

Opening Design and Production
Though central to the ideation of building information modelling (BIM), the vision of a shared collaborative environment for generating and managing building information stands only halfway fulfilled. In recent years, there has been a growing trend towards adopting open-source BIM software, as commercial software providers cannot keep pace with the increasingly complex requirements of their users. Open source, a transformative force in various domains, is embraced for its potential to stimulate innovation and collaboration while promoting cost efficiency. For BIM, open source can address its biggest hurdles – the lack of standardisation and interoperability issues. The need to negotiate between various disciplines – on the level of file formats, database architectures and semantic representations – poses fundamental questions on knowledge representation and its embedding.[7] Only through a common data framework will we be able to establish a consistent data foundation for future automation and machine-learning processes.[8] Yet if we are to build up domain knowledge as a community, it will not be enough to share the data, use the same data schemas or agree on open-data formats; we need also to start discussing and sharing software that governs the data: that generates, modifies, analyses, visualises or queries information on the data model. The oversight of knowledge embedded in software extends even to research foundation institutions, which, despite following open science principles, may overlook the need for a software management plan alongside a data management plan. Without clear guidelines, researchers and practitioners are left to devise their own strategies for storing, sharing and collaborating on the software they create.

Beyond current research efforts, we need to start looking at the production of digital tools that enable architecture in the same way as we look at the production of the end result – the building itself. Principles of reducing, reusing and recycling, which have become synonymous with sustainable practices, need to be extrapolated to the production and utilisation of digital tools with existing concepts such as permissive licensing, 'don't repeat yourself' (DRY) and incentives for upstreaming and refactoring. While creating public libraries is already the

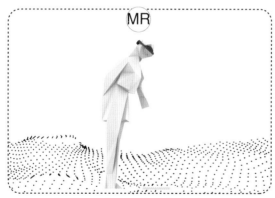

A: Creation of a digital twin

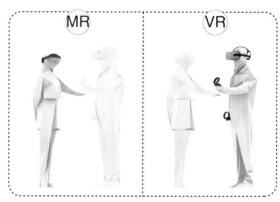

B: Localisation and meeting in virtual space

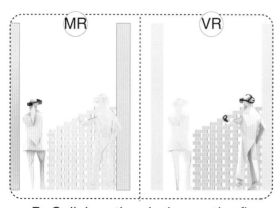

D: Collaborative design on-the-fly

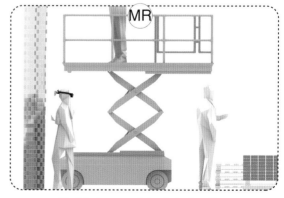

E: Holographic fabrication

first step towards software reuse, the community must find and promote best software practices applicable to the architecture domain. This includes developing application programming interface (API) specifications and documentation that is readable to the target audience, as starting points in this endeavour. To avoid redundant efforts and transcend individual endeavours, however, we need to foster collaborative approaches across software ecosystems, thereby accelerating the development of a robust software culture in architecture.

Here is where interdisciplinary experiments in digital fabrication – where the concept of interoperability has been replaced by a broader look at the infrastructures required to bridge software and hardware commanding both design and construction – can make a case for opening up knowledge in the architecture, engineering and construction industry. When design and fabrication are explicitly linked, their value chains, know-how, expertise and methods are co-dependent. Perhaps, over time, digital fabrication will emerge as architecture's justification to make design software readable, architectural knowledge shareable and our design decisions traceable. △

Notes

1. Jelle Feringa, 'The Promotion of the Architectural Model', in Andri Gerber and Brent Patterson (eds), *Metaphors in Architecture and Urbanism: An Introduction*, transcript Verlag (Bielefeld), 2013, pp 185–200.
2. Inés Ariza et al, 'Lost and Bound: Adaptive Detailing with Robotic Additive Joining for Reclaimed Steel', in *Beyond Optimization: Robotic Fabrication in Architecture, Art and Design*, Springer International Publishing (Cham), forthcoming.
3. See Romana Rust et al, 'Computational Design and Evaluation of Acoustic Diffusion Panels for the Immersive Design Lab: An Acoustic Design Case Study', in *Towards a New, Configurable Architecture: Proceedings of the 39th eCAADe Conference*, GRID (Novi Sad), 2021, pp 515–24.
4. Romana Rust et al, 'Spatial Wire Cutting: Cooperative Robotic Cutting of Non-Ruled Surface Geometries for Bespoke Building Components', in Sheng-Fen Chien et al (eds), *Proceedings of the 21st International Conference on Computer-Aided Architectural Design Research in Asia: Living Systems and Micro-Utopias: Towards Continuous Designing (CAADRIA 2016)*, CAADRIA (Hong Kong), 2016, pp 529–38.
5. https://compas.dev.
6. Daniela Mitterberger et al, 'Extended Reality Collaboration: Virtual and Mixed Reality System for Collaborative Design and Holographic-Assisted On-Site Fabrication', in Christoph Gengnagel et al (eds), *Towards Radical Regeneration: Design Modelling Symposium Berlin 2022*, Springer International Publishing (Cham), 2023, pp 283–95.
7. Alessio Lombardi, 'Interoperability Challenges: Exploring Trends, Patterns, Practices and Possible Futures for Enhanced Collaboration and Efficiency in the AEC Industry', in Pierpaolo Ruttico (ed), *Coding Architecture: Designing Toolkits, Workflows, Industry*, Springer Nature (Cham), 2024, pp 49–72.
8. 'Future AEC Software Specification / Data Framework': https://future-aec-software-specification.com/data-framework.

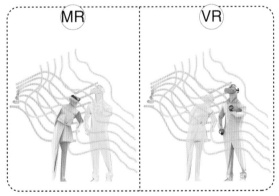

C: 3D sketching and annotating

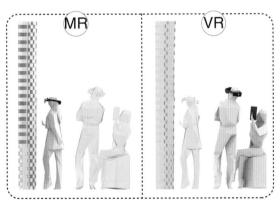

F: Fabrication supervision

> Perhaps, over time, digital fabrication will emerge as architecture's justification to make design software readable, architectural knowledge shareable and our design decisions traceable

Daniela Mitterberger / Gramazio Kohler Research,
Extended Reality Collaboration,
ETH Zurich,
2022

Phases of immersive decision-making include the creation of a digital twin, progressing to virtual meetings, collaborative on-the-fly design, and culminating in fabrication guided by holographic instructions and supervision.

Text © 2024 John Wiley & Sons Ltd. Images: pp 98–9 © Gramazio Kohler Research, ETH Zurich, photo Michael Lyrenmann; p 100 © Gramazio Kohler Research, Circular Engineering for Architecture and Chair of Steel and Composite Structures, ETH Zurich, photo Inés Ariza; p 101 © Gramazio Kohler Research, ETH Zurich, photo Gerhard Bliedung; pp 102, 104–7 © Gramazio Kohler Research, ETH Zurich; pp 103 © Gramazio Kohler Research, ETH Zurich, photo Norman Hack

Rethinking Digital
A collaborative Machines and

Daniela Mitterberger
and Kathrin Dörfler

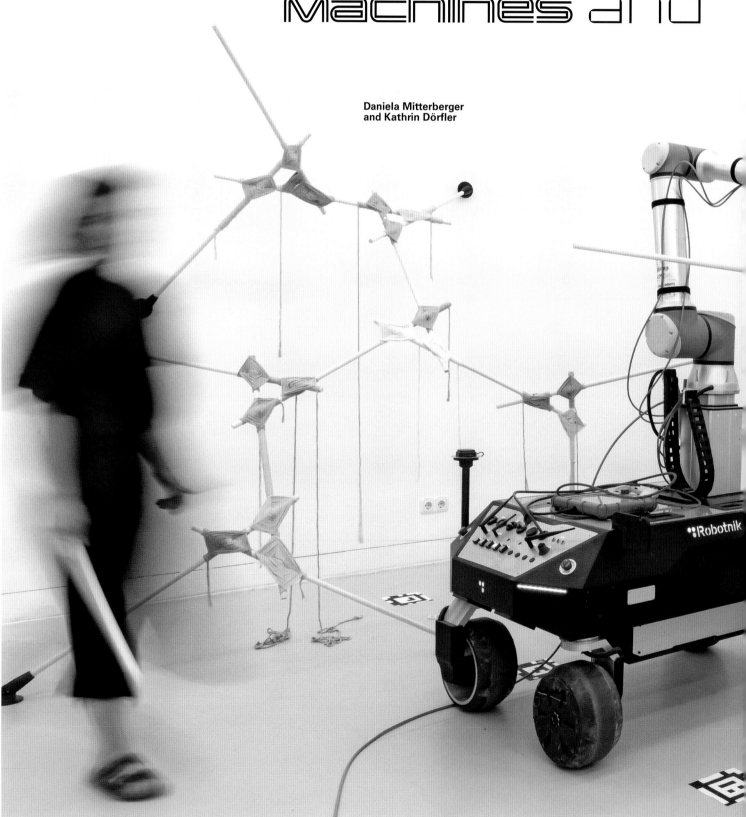

construction
future of humans,
craft

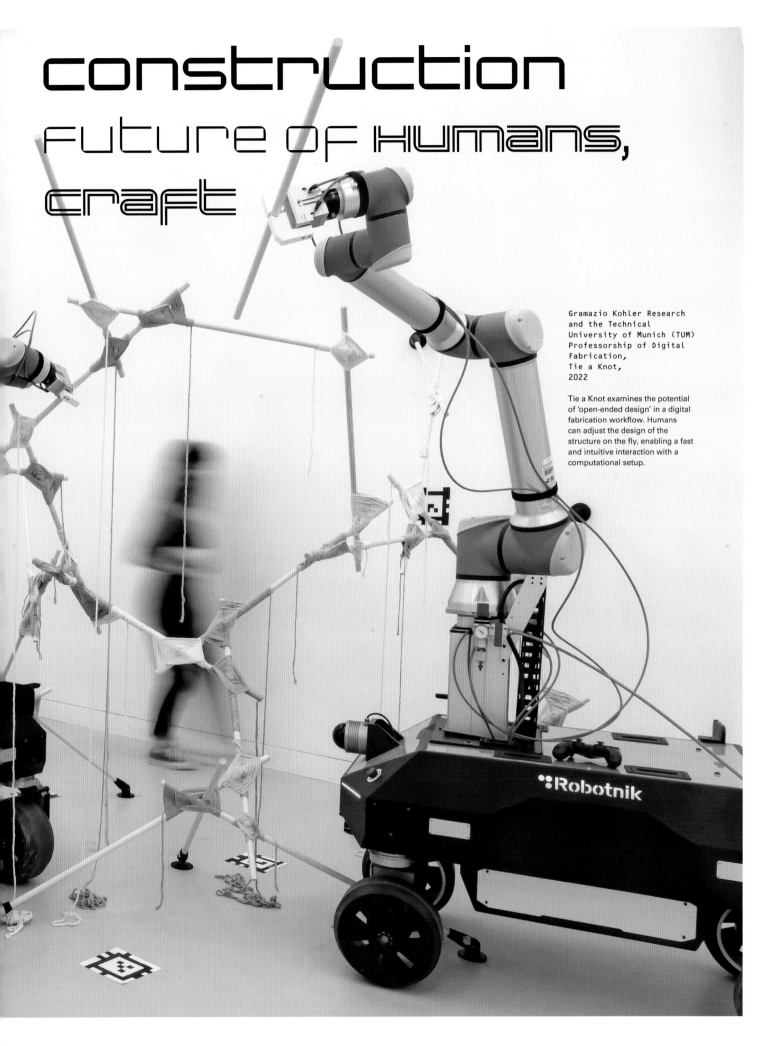

Gramazio Kohler Research
and the Technical
University of Munich (TUM)
Professorship of Digital
Fabrication,
Tie a Knot,
2022

Tie a Knot examines the potential of 'open-ended design' in a digital fabrication workflow. Humans can adjust the design of the structure on the fly, enabling a fast and intuitive interaction with a computational setup.

Gramazio Kohler Research,
Augmented Bricklaying,
Kitrvs winery,
Kitros, Greece,
2019

above: The geometry of the façade evokes the illusion of three-dimensional waves, computationally designed to gracefully symbolise the fluidity of wine, as well as mirroring the meandering contours of the nearby valley. Every brick, individually positioned and rotated, owes its precision to the augmented-reality (AR) system, enabling bricklayers to execute their craft with high accuracy and differentiation.

right: A camera sensor is capable of registering and tracking the exact location of bricks in 3D space. This spatial information is directly translated into craft-specific visual instructions – in this case, where and with how much force to tap a brick into its exact location.

The vast majority of construction industry processes are labour intensive, and often fraught with errors due to non-exact assembly tolerances and dimensions. Digital fabrication can mitigate or remove the potential of these anomalies occurring by negating or reducing human site-participation. But there is another side to the digital fabrication coin – the transition to more automated processes will liberate humans to become more collaborative and creative with machines both virtual and actual. **Daniela Mitterberger**, an Assistant Professor at Princeton University School of Architecture, and **Kathrin Dörfler**, Assistant Professor of Digital Fabrication at the Department of Architecture at the Technical University of Munich, highlight the evolving advantages of these new relationships.

The construction sector is on the verge of a transformation, driven by the possibilities of digital fabrication. They create aspirations for a future where fully automated construction processes replace manual labour, enhancing precision and quality in construction workflows. Yet these aspirations often overlook the technology's potential to empower human skills and foster meaningful collaboration between humans and machines.

Conventional narratives surrounding human–machine interaction often operate under the assumption that increasing machine autonomy leads to decreased human involvement. Research efforts frequently reflect this bias, aiming to minimise human intervention through ever more autonomous machines, but a compelling alternative also exists – a positive correlation between autonomy and collaboration. To achieve such a synergy, we require highly autonomous machines capable of comprehending and adapting to their surroundings.[1] Such robots, devices and algorithms are the foundation for interactive digital workflows that augment human actions and expand creative possibilities for shaping the built environment. By enabling machines to learn from the vast repository of human knowledge while fostering natural human–robot interaction, we could augment human skills and find novel and unexpected ways to seamlessly integrate human craft into digital construction processes.

A reimagined relationship between builders, materials, tools and technologies could help to realise what the interdisciplinary theorist María Puig de la Bellacasa calls technoscientific futurity,[2] which would facilitate a sustainable building future. This futurity does not overlook the contextual aspects of construction labour, physical exertion and cultural dependencies. Rather, in that setting, builders and craftspeople exemplify new ways of relating to materials, tools, buildings and architecture, that are enabled by emerging technology but that also cultivate the sociocultural dimensions of craft.

Presented here are four distinct architectural projects and experiments involving Gramazio Kohler Research and the Technical University of Munich (TUM) Professorship of Digital Fabrication, separately or in collaboration, all of which illustrate this new relationship between craft and digital construction technology, with each project exploring how to effectively centre builders and craftspeople within digitally supported building workflows. The projects consistently incentivise human engagement and present diverse opportunities for collaboration between human and computational intelligence.

Digitally Augmented Craftspeople

Craft, in its essence, is the skilful manual creation of objects, embodying artistic creativity and personal engagement. Craftspeople engage with physical materials, tools and processes, where tacit knowledge, an experiential understanding, is acquired and cultivated over long periods of time. This type of knowledge forms the core of what sociologist Richard Sennett discusses in his 2008 book *The Craftsman*, where he explores the concept of material awareness and how craftspeople gradually 'become the thing on which they are working '.[3] Merging craft sensibilities and practices with computational strategies and new making technologies therefore requires a fostering of active human engagement with the very materials in question.[4]

Augmented Bricklaying (2019) is a building project commissioned by Gramazio Kohler Research, in which the traditional craft of masonry was digitally enhanced for the assembly of the intricate, geometrically differentiated brickwork façade of the Kitrvs winery in Greece.[5] The project employed an ecosystem of digital tools, including visual sensors and augmented-reality (AR) interfaces, to elevate the expertise of masons in brickwork. A custom-developed AR guidance system enabled local Greek bricklayers to carry out a millimetre-precise assembly of 225 square metres (2,422 square feet) of fair-faced brick façade according to a 3D model, consisting of 13,596 individually rotated and tilted bricks. These digital tools inherently build on traditional masonry skills and principles, particularly the tacit knowledge of handling mortar and mortar joints, the dexterity required for bricklaying, and the flexibility and adaptability to handle malleable materials. In return, the intricate human craft skills are enhanced by the AR visual guidance system providing real-time three-dimensional accuracy and a direct link to the digital model, guiding the brick-and-mortar placement. To date, it is the largest project that has been built on site using an AR interface, merging digital technology with traditional artisanship in an entirely new construction process.

The Augmented Bricklaying guidance system combines a sensor system integrating an object-based tracking feature with a visualisation setup providing a craft-specific user interface. The sensor system facilitated highly accurate real-time tracking of individual objects, specifically bricks, in three-dimensional space.[6] Leveraging this spatial information – the precise locations of the bricks – the craft-specific user interface provided detailed visual guidance for specific masonry actions in real time. This involved guiding various bricklaying activities, such as carefully knocking the brick into the mortar at intended precise locations or applying mortar, using real-time visual cues derived from a digital model.

These dynamically generated instructions enabled the bricklayers to intuitively understand where to place the bricks in 3D space in correlation with a computer model.

The augmentation system was developed in collaboration with experienced bricklayers and proved exceptionally interactive and intuitive after an initial learning phase. When used in the real construction-site environment, it seamlessly establishes a direct link between the digital design environment, the skilled masons and the digital construction process. As a result, the system and the architectural outcome of this façade achieved a level of precision that surpasses conventional AR representations and is comparable to production by digitally controlled machines, but with a distinct and unmistakable human touch.

New Robot Collaborators

Along with elevating skills, there is also a need to mitigate physical challenges for builders and fabricators while keeping them cognitively engaged. With this concern in mind, the TUM Professorship of Digital Fabrication project Diversifying Construction (2023) looked into enhancing craftspeople through robots, exploring how robots can best support and collaborate with craftspeople skilled in masonry and bricklaying in the process of brickwork construction.[7] The aim was to have craftspeople and robots share tasks, combining their complementary abilities – human cognitive skills, versatility, fine motor skills and the robot's precision and endurance. The research further explored how such a sensible task distribution could contribute to the physical relief of construction labourers while retaining their physical engagement, tactile sense of the physical construction environment, decision-making and agency.

Designed to work safely alongside humans in shared workspaces, collaborative robots are built with lightweight materials, designed with rounded shapes and incorporate sensory systems for measuring and limiting force when needed. For this research project, a distinct workflow was set up between a skilled mason and a collaborative robot, intended to take turns performing physical tasks necessary to construct a brickwork structure. The bricks are assembled into their precise locations according to a given digital model by the robot, relieving the mason from measurement operations and repetitive lifting and assembly tasks. Manual application is used for the mortar, the malleable material that ensures the brick bond, accommodates material inaccuracies and enables the force transmission between the bricks. The bricklayer places the mortar beds, gently knocks bricks into them, cleans the mortar joints and triggers successive robot tasks.

left: The bricklayer is checking whether the brick is levelled correctly after the robot has placed it. Here, the transdisciplinary research approach – the testing and evaluation procedures with professionals – generated a spectrum of imaginations for how a collaborative robot might, or might not, be appropriately used, and whether or not it proves useful in certain tasks and scenarios.

```
Technical University of Munich (TUM)
Professorship of Digital Fabrication,
Diversifying Construction,
2023
```
left: The workflow tested with a professional bricklayer was concerned with the deliberate distribution of tasks, with the collaborating robot placing bricks based on a digital design model, and the bricklayer handling mortar application, refining brick placement, cleaning mortar joints and triggering robot tasks.

Gramazio Kohler Research and the
Technical University of Munich (TUM)
Professorship of Digital Fabrication,
Tie a Knot,
2022
above: The Tie a Knot project provides a cooperative workspace for two mobile robots and two humans. While robots are used to place precise elements and stabilise the structure, humans are used to tie rope joints and assemble intricate wooden elements.

Gramazio Kohler Research and the
Technical University of Munich (TUM)
Professorship of Digital Fabrication,
Collaborative Augmented Assembly,
2023
opposite: The project uses a phone-based augmented-reality app to enable intuitive task sharing between humans and robots.

The Human Element in Hybrid Construction
Preserving the multifaceted nuances of construction also involves integrating low-tech manual operations with high-tech digital fabrication processes – a concept examined through the collaborative project Tie a Knot (2022).[8] This project provides a cooperative workflow in which two mobile robots and two humans collectively assemble an intricate timber structure within a shared digital-physical workspace. The design encompassed a spatial reciprocal frame structure made of cylindrical wooden rods positioned in 3D space, supported only by mutual interlocking and tied together by knots with a soft rope.

In this context, robots carry out the precise placement and temporary stabilisation of the timber rods, crucial for high spatial accuracy. Simultaneously, humans undertake the intricate crafting of traditional rope joints, which demands a high level of dexterity.

Increasing creative engagement, humans were able to initiate assembly cycles, and these cycles were subsequently brought to completion with the support of robots. The system, set up with utmost flexibility and adaptability, allows humans to manually position timber rods at locations of their choice without a predefined design. Manually placed elements are digitally recorded and integrated into the digital model, based on which robot actions are generated adaptively. As such, the experimental study presents a fascinating human–machine cooperation, facilitating creative design choices and the act of manually joining elements as part of robotic assembly procedures. Here is where the exploration goes beyond mere technical functionalities, investigating pathways into synergistic human–robot cooperation and thereby offering valuable insights for the future integration of such hybrid workflows in the construction domain.

Building upon this concept, another collaborative Gramazio Kohler – TUM project, Collaborative Augmented Assembly (2023), introduces a phone-based mobile AR for humans to interface and seamlessly communicate with their robot collaborators during building construction. The custom-made AR app offers a simple interface for humans, allowing for real-time interaction between a 3D design model and the collaborating robots. It relies on cloud-based communication for information exchange among various mobile devices and robots. As the assembly unfolds, humans can utilise the AR app to preview robot motions, initiate robot actions and obtain detailed and spatially precise assembly instructions for the manual placement of timber rods and the placement of their mechanical connectors. Here again, robotic precision is utilised to place timber rods at strategically important locations. Human workers also play a crucial role, manually placing metal joints that require tightening, which adds a tactile dimension to the assembly process.

Gramazio Kohler Research and the Technical University of Munich (TUM) Professorship of Digital Fabrication, Collaborative Augmented Assembly, 2023
Collaborative Augmented Assembly supports cooperative human–robot fabrication of self-supporting structures.

Collaborative Narratives for Digital Construction

The outcomes of these experiments suggest that there may be room for achieving new processes of collaboration within the bounds of automation narratives for digital construction. While certain aspects of these workflows, such as flexible task distribution and communication fluidity, will require further development and in-depth research, such technological advancements could effectively be harnessed to facilitate future collaborations between humans and machines that improve working conditions, foster human engagement and increase the interactivity of machines. Such collaborations can preserve the multifaceted nuances of construction labour, which encompass cultural heritage, knowledge preservation, human–tool relationships and various individualised craft skills influenced by geographical context, class or gender, while capitalising on technological possibilities such as skill elevation, increased precision, improved repeatability or reduced physical exertion. ᗡ

Notes
1. See Jenay M Beer, Arthur Fisk and Wendy Anne Rogers, 'Toward a Framework for Levels of Robot Autonomy in Human-Robot Interaction', *Journal of Human-Robot Interaction* 3 (2), 2014, pp 74–99.
2. See María Puig de la Bellacasa, *Matters of Care: Speculative Ethics in More than Human Worlds*, University of Minnesota Press (Minneapolis, MN), 2017.
3. See Richard Sennett, *The Craftsman*, Yale University Press (New Haven, CT), 2008, p 174.
4. See Terry Knight, 'Craft, Performance, and Grammars', in Ji-Hyun Lee (ed), *Computational Studies on Cultural Variation and Heredity*, Springer (Singapore), 2018, pp 205–24.
5. See Daniela Mitterberger *et al*, 'Augmented Bricklaying: Human–Machine Interaction for In Situ Assembly of Complex Brickwork using Object-Aware Augmented Reality', *Construction Robotics* 4 (3–4), 2020, pp 151–61.
6. See Timothy Sandy and Jonas Buchli, 'Object-Based Visual-Inertial Tracking for Additive Fabrication' *IEEE Robotics and Automation Letters* 3 (3), 2018, pp 1,370–77.
7. See Lidia Atanasova and Kathrin Dörfler, 'Diversifying Construction', in Anh-Link Ngo *et al* (eds), *ARCH+: The Great Repair – Politics of the Repair Society*, ARCH+ (Berlin), 2023, pp 170–71.
8. See Daniela Mitterberger *et al*, 'Tie a Knot: Human–Robot Cooperative Workflow for Assembling Wooden Structures using Rope Joints', *Construction Robotics* 6 (3–4), 2022, pp 277–92.

Text © 2024 John Wiley & Sons Ltd. Images: pp 108–9, 114–17© Gramazio Kohler Research and TUM Professorship of Digital Fabrication; p 110 © Gramazio Kohler Research, ETH Zurich; pp 112–13 © TUM Professorship of Digital Fabrication

Selen Ercan Jenny and Abel Gawel

Sustainable Through Robotic Past and

Construction On-site Fabrication Future Concepts

Gramazio Kohler Research in collaboration
with the Robotic Systems Lab,
Autonomous Systems Lab and Chair of
Geosensors and Engineering Geodesy,
Robotic Plaster Spraying (RPS),
ETH Zurich,
2023
RPS in 1:1 scale – articulated plasterwork produced
on site, testing the applicability and scalability of the
robotic process.

The use of robotics in construction has been experimented with for nearly fifty years. **Selen Ercan Jenny**, Guest-Editor of this ⌀, and **Abel Gawel**, Principal Researcher in Computer Vision and Machine Learning with the Huawei European Research Center, run us through a brief history of such developments, and describe the sustainable advantages of contemporary research in this field alongside the many hurdles still to be overcome to achieve the full adoption of robots on construction sites.

Gramazio Kohler Research,
ECHORD dimRob,
ETH Zurich,
2011

left: The dimRob robotic platform is an industrial robot (ABB IRB 4600) mounted on a compact mobile track system. Project team: Volker Helm, Ralph Bärtschi, Tobias Bonwetsch, Selen Ercan, Ryan Luke Johns and Dominik Weber. Industry partner: Bachmann Engineering.

opposite: A timber modular wall, fabricated by the mobile robotic platform of dimRob.

Since the 1980s, studies have sought to identify robust, productive and affordable approaches to a vast range of fields through the use of robotic arms: from the production of brick walls, to assembly and finishing operations, surveying and positioning tasks, inspection, repair and maintenance activities, material handling, concrete placement and on-site finishing operations.[1] Concurrently, industry, led by Japanese corporations, has made inroads into the automation of existing construction processes using on-site machines,[2] although these have yet to bring autonomy, versatility or sustainability to the processes. Rapid advances in autonomous mobile robots research[3] in the first decades of the 2000s led to the emergence of intelligent and increasingly versatile construction robots. Of particular note has been autonomous reality capture, as a maturing, commercially available application,[4] and there is a growing interest within the architectural field in the connection between sustainable materials, novel fabrication processes and robotic technologies.

In 1987, the UN Brundtland Commission defined sustainability as 'meeting the needs of the present without compromising the ability of future generations to meet their own needs'.[5] Among the UN's 17 Sustainable Development Goals (SDGs), several are relevant for the construction industry today: seeking to ensure healthy lives and to contribute to the wellbeing of people of all ages; providing full, productive and decent employment for all; encouraging innovation for the creation of resilient infrastructure; and taking urgent action to combat climate change and its impacts. Relating to some of these goals, the four research projects below, all of which were developed at ETH Zurich over a 10-year period, propose fabrication methods and concepts that make use of on-site machines in the architectural field to increase accuracy, robustness, productivity, material savings and workplace safety.

Mobility, Man–Machine Interaction and Dimensional Tolerance Handling

The ECHORD (European Clearing House for Open Robotics Development) dimRob project (2011–12) conducted by Gramazio Kohler Research is a first step in the evolution of mobile robotics on construction sites, building upon innovative paradigms in man–machine interaction to address the lack of precision and the large deviations commonly faced in this context. Complementing those, dimRob's key features include its implementation of a cognitive strategy for self-positioning of the mobile robotic platform. This challenge is addressed by 3D scanning of local reference points in the workspace, which the mobile platform then follows when repositioning itself – facilitating the fabrication process and supporting the creation of large architectural structures in several segments, such as the 8-metre-long (26-foot) timber modular wall that was the project's first architectural-scale experiment.

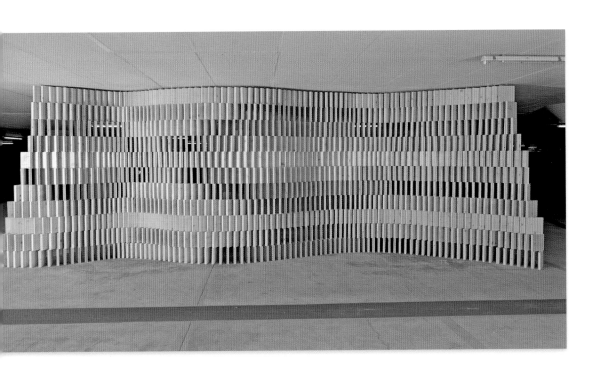

Gramazio Kohler Research in collaboration with ROB Technologies,
Mobile Robotic Tiling,
Singapore-ETH Centre,
Singapore,
2014

right: The Mobile Robotic Tiling machine on site, safely alongside human operators, promising productivity gains through robotic automation. Project team: Selen Ercan Jenny, Jonas Furrer and David Jenny. Industry partners and sponsors: SIKA Technology, LCS Optiroc and Bouygues Construction.

ETH Autonomous Systems Lab and Robotics Systems Lab in collaboration with Hilti,
Waco,
ETH Zurich,
2021

opposite left: A custom autonomous mobile manipulator robot (Waco) with a grinding tool from manufacturer Hilti.

Productivity Increase for Construction

Robotic technologies have the potential to dramatically increase productivity in the building sector while also maintaining high construction quality. Tiling has traditionally been a labour-intensive, slow, manual process that is prone to errors, and workmanship quality decreases as skilled labour becomes scarce – a situation that is expected to worsen over the coming decade.

The Mobile Robotic Tiling project (2013–16), conducted by Gramazio Kohler Research and ROB Technologies at the Singapore-ETH Centre, proposed a design for an on-site robotic tiling machine within the high-rise context of Singapore that can deliver high accuracy at more than twice the speed of conventional processes, acting as a robotic co-worker safely alongside a human operator without requiring particular security measures. The system comprises a versatile robotic arm mounted on a mobile platform and a custom-made end-effector that allows the picking and placement of a tile, along with a specialised nozzle tool that applies the adhesive. It removes the need to change tools between adhesive application and tile placement, and is designed to operate in constrained spaces.

The project envisioned on-site robotic tiling to be just the start. In a wider plan to expand throughout the region, the ultimate goal entailed creating an internationally recognised Swiss technology to be adopted globally, transforming the way we think about construction work and the way we build cities.

Waco: High-Accuracy Mobile Manipulation

In contrast to dimRob and Mobile Robotic Tiling, Waco is a multi-purpose mobile manipulator robot that can execute different construction tasks autonomously, including setting out, drilling and grinding tasks. Developed by a multidisciplinary team at ETH Zurich and power tool developer and manufacturer Hilti between 2018 and 2022, it comprises a mobile base, a robotic arm and an adaptable tool mount, and is equipped with a range of sensors for navigation and, optionally, for high-accuracy task execution (cameras, structured lights, range meters). The project envisaged the autonomous execution of efficient, resilient workflows derived from building models, using appropriate references for the on-site localisation of the robot to perform high-accuracy tasks, and feeding execution results back to a building model.[6]

The studies carried out for the Waco project investigated the interrelationships between as-planned building information and as-built robotic sensing.[7] They revealed the real-time link between the two to be crucial for autonomous construction operations, for the well-defined handling of reference points and deviations that can be directly derived from the building model.[8] One important gap that Waco fills lies between mobile robotic sensing, which is not fit to deliver the accuracies required in construction processes, and accurate survey-grade sensing.

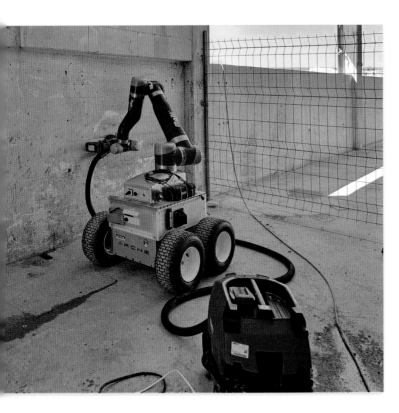

Gramazio Kohler Research in collaboration with ROB Technologies,
Mobile Robotic Tiling,
Singapore–ETH Centre, Singapore,
2014

below: A custom end-effector, comprising a main gripping tool and an automated adhesive applicator – the first-generation design of a special nozzle and extrusion tool for the Mobile Robotic Tiling machine.

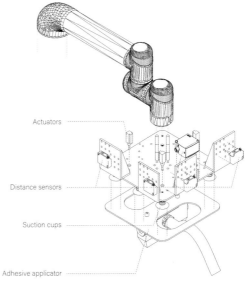

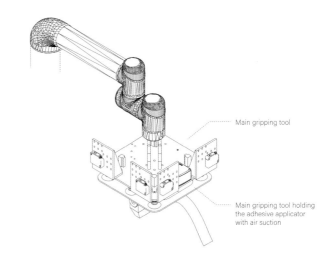

Waco combines the two different realms of mobile robotic sensing and construction technology. The sensing paradigm brings as-designed building information together with the as-built environment and addresses the challenges of as-built sensing performance related to different environments and job sites, helping minimise construction errors. An exciting emerging research direction in this regard is self-supervised adaptation of the sensing to individual construction scenes, such as allowing a robot to autonomously adjust its sensing parameters to specific situations using artificial intelligence (AI) concept continual learning (CL) mechanisms,[9] where a model continuously learns from new data without forgetting previously learned information. Through this approach, AI systems are able to adapt to new tasks and information over time, making the technology more versatile and efficient at handling evolving real-world scenarios.

As a research platform, the technical design of Waco is open-sourced, fostering innovation and strengthening partnerships within different fields.[10] Considerable effort has been put into establishing a construction robotics focus group at ETH Zurich, merging architecture, engineering and construction research with mobile robotics, with a view to identifying synergies and potential hurdles and ultimately bringing the benefits of digital construction on site.

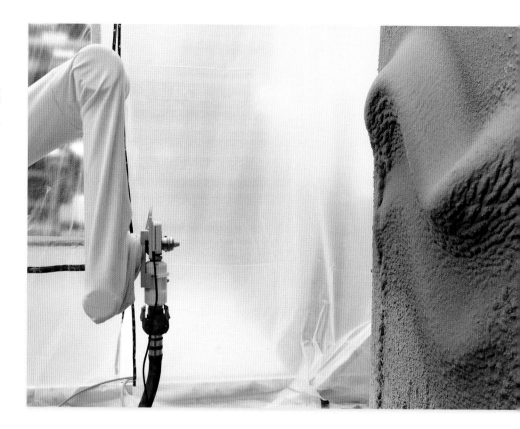

Gramazio Kohler Research in collaboration with the Robotic Systems Lab, Autonomous Systems Lab and Chair of Geosensors and Engineering Geodesy,
Robotic Plaster Spraying (RPS), ETH Zurich,
2020

left: The novelty of RPS lies in the proposed gravity-resisting printing technique.

right: RPS revisits customised 3D plasterwork in architecture, produced with minimal waste. This sample was produced without formwork, support structures or additional tools, using only the proposed, additive-only adaptive thin-layer printing technique. Project team: Selen Ercan Jenny, Ena Lloret-Fritschi, Eliott Sounigo, Ping-Hsun Tsai, Valens Frangez, Philippe Fleischmann and Luca Ebner. Industry partners and sponsors: Hilti and Giovanni Russo.

Material Savings with a Digital Craft

As exemplified by the dimRob, Mobile Robotic Tiling and Waco projects, robotic systems can dramatically increase not just productivity, but also accuracy and robustness in the building sector. Furthermore, the digital control of construction tasks opens up opportunities for bespoke architectural designs due to its easy adaptation to unique requirements, while also contributing to reduced construction mistakes and on-site material savings.

In contemporary construction practice, cementitious plaster is commonly used to protect and stabilise buildings, making them more durable, although it is a labour-intensive process that involves specialised skills. Unfortunately, contemporary plastering processes make use of subtractive – and therefore waste-generating – fabrication techniques, and can often only be used to produce flat, standardised plasterwork, relying on the removal and discarding of significant quantities of the wet material to shape, smoothen or level the surface.

Robotic Plaster Spraying (RPS) is an innovative additive-only, spray-based plaster printing technique that has been undergoing development by an interdisciplinary team at ETH Zurich since 2018 and allows the production of standard and custom plastered surfaces with minimal waste. The project revisits bespoke 3D plasterwork in architecture that can lead to structural, acoustic or thermal optimisation. It introduces new degrees of design freedom to the craft of plastering through digitisation. Instead of spraying centimetre-thick layers, it involves a thin-layer, high-resolution, spray-based printing technique. The applicability and scalability of this technique have been explored through empirical research with architectural prototypes in which multiple, millimetre-thin layers of plaster are sprayed onto a building structure, allowing the incremental build-up of 3D forms (for the plastic shaping of wet material) without the need for additional smoothing or profiling tools, formwork or support structures, and without any subtractive steps to remove material.

Currently, studies at ETH Zurich are continuing to validate the robustness, scalability, applicability and performance of the developed technology, and there is a long-term plan in place to demonstrate the viability of sustainable automated construction, efficiency in sectors that directly impact society, safety on job sites and lowered material use.

Towards Sustainable On-site Fabrication

Starting from a technological perspective, dimRob has taken the first step towards developing a robust system that takes full advantage of the complementary strengths of humans and machines on site, bringing the speed and precision of automation, and fostering innovation. On the socio-economic side, people prefer cleaner, less strenuous, better-paid jobs; and, as a co-worker taking on the strenuous part of the task, Mobile Robotic Tiling can be considered a possible solution to the creation of safer and healthier on-site working environments, promoting wellbeing for all. At a societal and environmental level, RPS takes a novel approach to a craft that is as old as the history of building, in terms of material use and the necessary formwork and tools,

also promising a reduction in material waste through a digitally controlled, additive-only process with contributions to sustainable consumption and production patterns.

Given the high degree of digitisation that can now be found on contemporary construction sites, construction processes are analytically well described, and are an ideal application scenario, as semi-structured environments, for autonomous mobile robotic systems from a roboticist point of view, as shown by Waco. The advances in robotic hardware, sensing and machine intelligence facilitate efficient and autonomous micro-controlled construction processes, accurate documentation and holistic process integration without analogue gaps, as well as low error rates, all of which could contribute to building the resilient infrastructures of the future.

That said, several issues remain as hurdles to bringing scientific proof-of-concept construction robots into daily operation. Their performance can vary greatly in different environments, and there are myriad situations that cannot be anticipated. Mechanisms that allow robots to deal with scene semantics are needed, along with continuous adaptations to situations that human workers, when encountering them, are able to resolve by drawing upon experience or intuition.

Until these fundamental challenges are overcome, for on-site robotic fabrication systems to be of practical use in the construction sector, they must integrate the cognitive skills of human workers in an efficient way, and be productive, while addressing material savings and environmental values. It is clear that automated robotic technologies on construction sites – or any such process, including conventional construction methods – have to be proven to be environmentally advantageous, in addition to their economic and technological feasibility. We are now at a turning point in exploring how on-site machines could help to meet the sustainability goals of the near future. 🗈

Notes
1. Takatoshi Ueno, 'Trend Analysis of ISARC (International Symposium on Automation and Robotics in Construction)', in *Proceedings of the 15th ISARC International Symposium on Automation and Robotics in Construction*, 1998, pp 5–12.
2. See Mark Taylor, Sam Wamuziri and Ian Smith, 'Automated Construction in Japan', in *Proceedings of the Institution of Civil Engineers-Civil Engineering* 156 (1), 2003, pp 34–41.
3. See Roland Siegwart, Illah Reza Nourbakhsh and Davide Scaramuzza, *Introduction to Autonomous Mobile Robots*, MIT Press (Cambridge, MA), 2011.
4. See HS Xie, I Brilakis and E Loscos, 'Reality Capture: Photography, Videos, Laser Scanning and Drones', in Marzia Bolpagni, Rui Gavina and Diogo Ribeiro (eds), *Industry 4.0 for the Built Environment: Methodologies, Technologies and Skills*, Springer (Cham), 2022, pp 443–69.
5. Gro Harlem Brundtland et al, *Report of the World Commission on Environment and Development: Our Common Future*, United Nations, 1987, section I.27: www.un.org/en/academic-impact/sustainability.
6. Abel Gawel et al, 'A Fully-Integrated Sensing and Control System for High-Accuracy Mobile Robotic Building Construction', in *2019 IEEE/RSJ International Conference on Intelligent Robots and Systems (IROS)*, 2019, pp 2300–2307.
7. Hermann Blum et al, 'Precise Robot Localization in Architectural 3D Plans', in *ISARC: Proceedings of the 38th International Symposium on Automation and Robotics in Construction*, 2021, pp 755–62.
8. Selen Ercan et al, 'Online Synchronization of Building Model for On-Site Mobile Robotic Construction', in *Proceedings of the 37th International Symposium on Automation and Robotics in Construction*, 2020, pp 1,508–14.
9. See Hermann Blum et al, 'Self-Improving Semantic Perception for Indoor Localization', in *5th Conference on Robot Learning (CoRL 2021)*, 2022, pp 1,211–22.
10. See ETHZ RobotX SuperMegaBot: https://ethz-robotx.github.io/SuperMegaBot/.

Text © 2024 John Wiley & Sons Ltd. Images: pp 118–22, 123(br), 124–5 © Gramazio Kohler Research, ETH Zurich; p 123(tl) © Autonomous Systems Lab, ETH Zurich

**Silke Langenberg,
Sarah M Schlachetzki
and Robin Rehm**

■ Walter Gropius,
Dessau-Törten housing estate
under construction,
Dessau, Germany,
1927/28
Slag concrete in the Bauhaus context:
infill walls are being lined with small slag
concrete slabs and cellular concrete.

Concrete Patents
Innovative Building Materials and Construction Processes

The world of patents has always expanded exponentially in line with innovative inventions and materials, and this has been particularly true in the first few decades of the 21st century, which have delivered numerous construction materials and jointing and bonding techniques. ETH Zurich's **Silke Langenberg, Sarah M Schlachetzki and Robin Rehm** draw attention to the fact that within the academic field of construction heritage, little research has been done to review historical patents for possible sustainable construction materials and methods left by the wayside in the pursuit of industrialised architectural modernism.

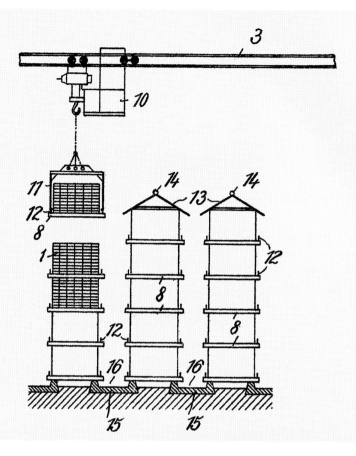

■ Johan van Item,
German Patent No 435,195 for storage options of freshly pressed slag stones,
1926
right: Mass-production of slag stones entailed new inventions for their drying and storage.

■ Rasmus A Larsen and Axel Nielsen,
Improvements in or relating to Building Structures,
British Patent No 1,001,759,
1965
opposite: The Danish company Larsen & Nielsen was highly successful in the 1950s and 1960s. Its patents for large slab constructions migrated across borders through the diverse licences that were issued.

The rather young 21st century presents us with a multitude of materials, constructions, building methods and processes that would not exist without the inventive spirit of previous generations of engineers and architects. By protecting inventions or ensuring a dominant market position for the inventors or companies submitting them, patents have played an important role since the end of the 19th century. In the course of the growing industrialisation of construction sites in the 20th century, their significance and scope grew considerably: initially granted for new construction machinery and building materials, patents increasingly came into question for new construction methods and products. This went hand in hand with the development of materials and binding agents specifically intended for construction, and concrete as a building material plays a decisive role in this context.

Globalisation and digitalisation have multiplied the number of patents in the construction sector. One reason for this increase in the last two decades has been the number of patent applications filed by architects seeking to protect their designs from imitation. Another is the number of new developments in digital fabrication, as well as the adoption and further development of earlier inventions. A growing number of projects and research carried out by leading universities in this field result in patent applications.

Digital fabrication presents opportunities to rethink existing inventions. At the same time, patents developed in times of crisis and scarcity could also help address today's pressing challenges in the face of climate change and resource depletion. It seems more meaningful, more timely and perhaps more exciting than ever to take a closer look at historical patents – both successful and forgotten ones. Under changing circumstances, they could unfold unexpected potential or simply inspire new ideas.

Upgrade of Waste Materials

In the first half of the 20th century, stones made from slag and ashes conveyed early attempts in what is discussed today as 'sustainable', modular construction. Buildings were broken down into parts in a modularity affecting their entire structure. Preproduction of their parts – artificial stones, in this case – was streamlined and rendered as efficient as possible. Mechanisation was central: all parts of the building process that could be standardised were serially mass-produced in advance. Assembly took place at the end. Efficient planning, automation and well-organised execution reduced the costs of building to fractions of older, more labour-intensive formats. The basic idea of turning data into things can be retraced to these early attempts to standardise building elements as well as the building process.

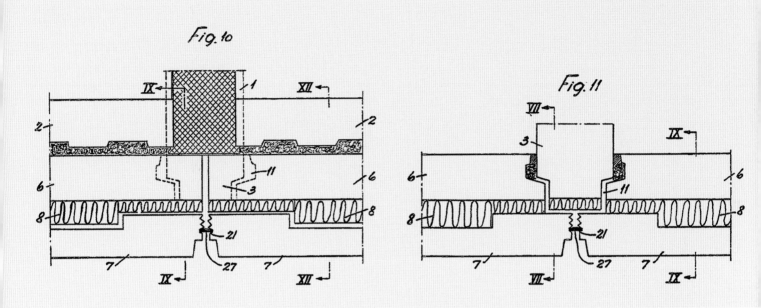

The hydraulic qualities of blast furnace slag – and to a lesser extent of regular slag and fly ash – had been discovered in the second half of the 19th century. It was in times of economic crisis after the First World War, however, that slag came into focus with novel thrust. In most countries in Europe, coal was scarce, which made energy expensive and regular bricks costly to produce. Like bricks, artificial stones made of slag and ashes existed in a limited number of different forms yet promised sheer endless reproduction. And the use of slag profited from the hydraulic qualities of a waste product: it reduced the amounts of slag otherwise deposited at space-consuming disposal sites, and saved energy. Foaming the slag produced lightweight stones of considerable logistical advantage. In short, slag stones surfaced anew as a sustainable alternative to clinker and concrete.[1]

Patent holders and patent seekers in the field of artificial stones diversified their strategies early on. They filed for patents pertaining to all parts of the production process. Patents issued concerned the production of the paste-like mixture of which the stones were made, machines amalgamating and processing this mixture, presses, and storage options of freshly fabricated stones. *In toto*, patents covered every aspect of cost reduction, efficiency and proto-automation, right up to potentially machine-driven devices in an increasingly mechanised building process.

Building With Systems
The reduction of disparate building elements was an essential prerequisite for industrial construction before digitalisation made individual mass-production possible. This was one of the main reasons behind the efforts towards system buildings in the 1950s and especially the 1960s in order to remedy the housing shortage that prevailed throughout Europe after the end of the Second World War.[2] In addition to simple skeletal systems filled with serially produced bricks or hollow blocks, mass housing construction now employed mainly solid, large precast slab systems, the standardised structural elements of which were prefabricated directly on the building site or in a nearby factory.[3] Igéco SA, founded in 1956, was one of Switzerland's most innovative industrial groups operating in this sector. The company served as a pioneer in the establishment of precast concrete slab construction and was a major driving force behind related developments. It operated using licences from the Danish construction company Larsen & Nielsen, such as the patent *Improvements in or relating to Building Structures* from 1965.[4]

After certain adaptations to Swiss conditions, Igéco constructed residential buildings from prefabricated concrete elements. Horizontal ceiling slabs were combined with load-bearing walls and facades were designed as sandwich panels with integrated insulation layers.[5] To prefabricate the large-format concrete elements in the workshop, the company developed its own vibrating tables with timber and steel moulds, for which it applied for the patent *Dispositif pour le moulage de panneaux en béton* in 1963. The patent describes how the concrete panels are cast in vertical battery moulds.[6] The elements were assembled on site using metal plate anchors, which Igéco had also legally protected with the patent *Panneau de béton et procédé de mise en œuvre de ce panneau*.[7] The metal plate anchors allow the panels to support themselves without the need for grouting. The La Gradelle development in Geneva, planned by Jean Hentsch and Jacques Zbinden between 1961 and 1963, is one of the best-known projects built using the Igéco construction system and demonstrates the many design variations it offers. With its licences and patents, Igéco had a considerable market share in Switzerland alongside Ernst Göhner AG, and was responsible for the construction of numerous apartments in the 1960s and 1970s. The developments described here for Switzerland can be observed to a much greater extent in the larger neighbouring countries affected by severe war damage, especially Germany and France.[8]

■ Jean Hentsch and Jacques Zbinden,
La Gradelle housing complex,
Geneva,
1963
opposite: Innovative slab construction far from monotonous satellite towns: one of the best-known projects using the Igéco construction system.

■ Igéco SA,
Dispositif pour le moulage de panneaux en béton (Device for modelling concrete panels),
Swiss Patent No 402,707,
1965
above: The Swiss company Igéco was a licensee of foreign patents and successfully applied for its own: in this case for a device for moulding concrete panels.

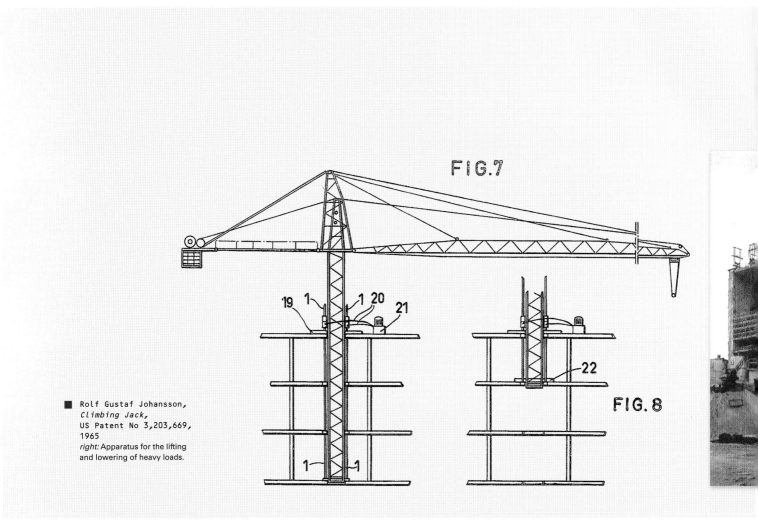

Rolf Gustaf Johansson,
Climbing Jack,
US Patent No 3,203,669,
1965
right: Apparatus for the lifting and lowering of heavy loads.

Concrete Processes

The standardisation of buildings and the reduction of disparate building elements significantly aided the industrialisation of construction sites. In the second half of the 20th century, concrete largely replaced the more expensive steel as the predominant building material, since the need to construct large volumes of buildings favoured the serial prefabrication of concrete elements.[9] This is also reflected in the increasing number of patents at the time for static and mechanical formwork systems used directly on construction sites, in addition to those for the development of specific connection details or support structures for the efficient joining and assembly of prefabricated elements. For the construction of high-rise buildings, staircase and elevator cores, climbing and sliding formworks were used, moved floor by floor or continuously after the concrete had dried out. These processes were also accompanied by various patents for special climbing winches and jacking machines.[10] Some of these methods were developed as early as the turn of the century for the construction of silos. The first patent for slipcasting, for example, was filed in 1900 by engineer Charles F Haglin.[11] The relevance of American industrial and silo buildings as models for the industrialisation of construction was described vividly in 1986 by Reyner Banham in *A Concrete Atlantis*.[12]

One of the now largely forgotten patents is the lift-slab method. With the goal of saving on formwork and scaffolding, the slabs of a multistorey building are cast on top of each other on the ground with separating layers and then lifted into their final position from top to bottom. Although this method was developed as early as 1913 and was tested near Paris in 1946,[13] it took until 1955 to be patented as *Apparatus for Erecting a Building* by Thomas B Slick in the US.[14] Both the highly successful slipform method and the rarely used lift-slab method have been rediscovered in the course of the digitalisation of construction processes or have been further developed in terms of their potential for sustainable construction. For instance, smart dynamic casting uses a robotic setup to control the slipform in terms of speed and geometry to enable the production of complex forms in a continuous concrete casting process.[15] Cast on cast, on the other hand, utilises the potential of the lift-slab method to save on complex formwork.[16]

■ The Regional Tax Office under construction, Münster, Germany, 1967
below: The lift-slab method, newly developed by Hochtief AG, employed at a construction site in Münster.

■ Thomas B Slick, *Apparatus for Erecting a Building,* US Patent No 2,715,013, 1955
right: Slick's invention included a jack and hoisting unit for more economical concrete construction.

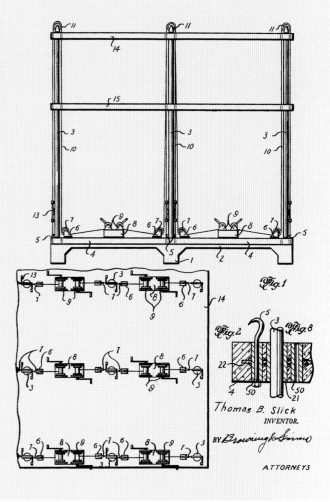

The development of new, sustainable materials and construction processes has so far progressed largely independent of research in the academic field of construction heritage. While most new technologies provide historical references to their direct predecessors, a systematic investigation of the potential of past inventions has yet to be carried out. Closer collaboration between the different research disciplines opens up new opportunities to develop the undiscovered potential of historical patents in architecture for making the past productive.[17] ⌂

Translated from the German by Orkun Kasap

Notes
1. See Arthur Guttmann, *Die Verwendung der Hochofenschlacke*, Verlag Stahleisen (Düsseldorf), 1934, and GW Josephson, F Sillers and DG Runner, *Iron Blast-Furnace Slag: Production, Processing, Properties, and Uses*, US Government Printing Office (Washington DC), 1949.
2. Barry Bergdoll and Peter Christensen (eds), *Home Delivery: Fabricating the Modern Dwelling*, Birkhäuser (Basel), 2008.
3. Silke Langenberg, 'Highly Visible and Highly Valuable: Big Housing Estates of the Boom Years', in Maren Harnack, Matalie Heger and Matthias Brunner (eds), *Adaptive Re-Use: Strategies for Post-War Modernist Housing*, Jovis (Berlin), 2020, pp 37–45.
4. Rasmus Andreas Larsen and Axel Nielsen, *Improvements in or relating to Building Structures*, British Patent No 1,001,759, London Patent Office, 1965.
5. Fanny Vuagniaux, 'Igéco', in ICOMOS Suisse (ed), *System & Serie: Systembau in der Schweiz – Geschichte und Erhaltung*, gta Verlag (Zurich), 2022, p 74.
6. Igéco SA, *Dispositif pour le moulage de panneaux en béton*, Swiss Patent No 402,707, Swiss Federal Institute of Intellectual Property, 1965.
7. Igéco SA, *Panneau de béton et procédé de mise en œuvre de ce panneau*, French Patent No 1,412,984, French National Institute of Industrial Property, 1965.
8. Thomas Schmid and Carlo Testa, *Systems Buildings / Bauen mit Systemen / Constructions modulaires*, Artemis (Zurich), 1969.
9. Silke Langenberg, 'The Hidden Potential of Building Systems: The Marburg Building System as an Example', in *Construction History, International Journal of the Construction History Society* 28 (2), 2013, pp 105–26.
10. Rolf Gustaf Johansson, *Climbing Jack*, US Patent No 3,203,669, US Patent and Trademark Office, 1965.
11. William J Brown, *American Colossus: The Grain Elevator, 1843 to 1943*, Colossal Books (New York), 2009.
12. Reyner Banham, *A Concrete Atlantis: US Industrial Building and European Modern Architecture*, MIT Press (Cambridge, MA), 1986.
13. Oskar Büttner, *Hubverfahren im Hochbau*, dva (Stuttgart), 1972, pp 71–82.
14. Thomas B Slick, *Apparatus for Erecting a Building*, US Patent No 2,715,013, US Patent and Trademark Office, 1955.
15. Ena Lloret-Fritschi, 'Smart Dynamic Casting: A Digital Fabrication Method for Non-standard Concrete Structures', doctoral thesis, ETH Zurich, 2016: https://doi.org/10.3929/ethz-a-010800371.
16. Lluís E Monzó, 'Design of CASTonCAST Shell Structures Based on Load Path Network Method', doctoral thesis, ETH Zurich, 2017: https://doi.org/10.3929/ethz-b-000274137.
17. Department of Architecture, ETH Zurich, 'Making the Past Productive', *Strategy*, 2023, p 15: https://arch.ethz.ch/en/departement/strategie.html.

Text © 2024 John Wiley & Sons Ltd. Images: p 126 Bauhaus-Archiv Berlin. © DACS 2024; p 130 © Oliver Marc Hänni; p 133(l) published in Oskar Büttner, *Hubverfahren im Hochbau*, VEB Verlag für Bauwesen (Berlin), 1972, p 84

FROM ANOTHER PERSPECTIVE *A Word from △D Editor Neil Spiller*

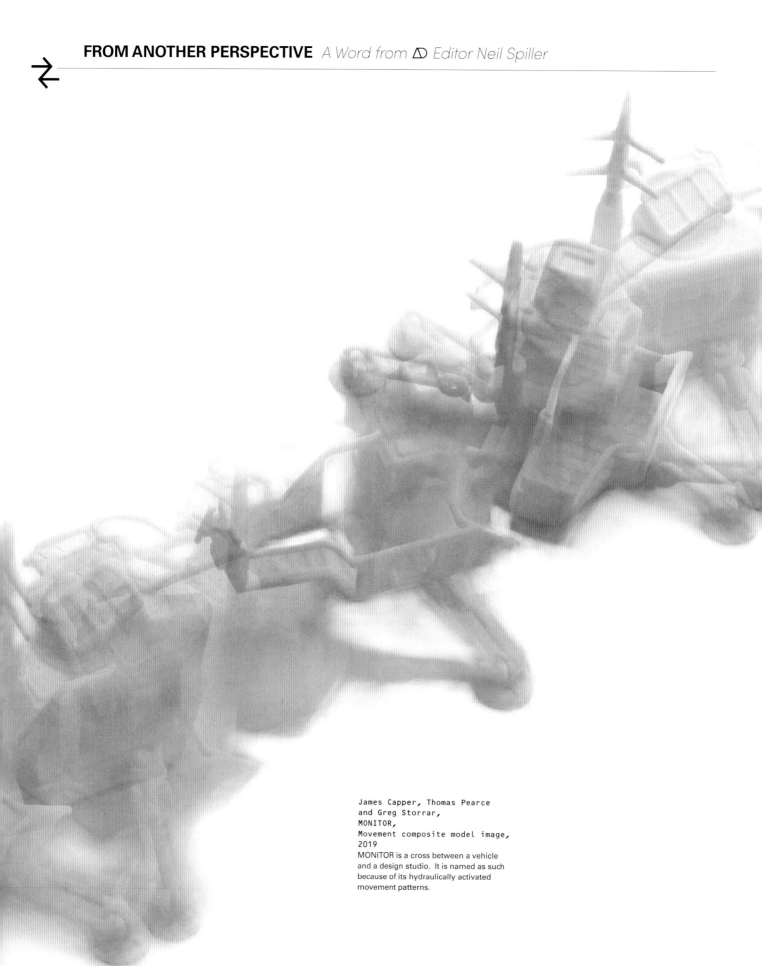

James Capper, Thomas Pearce
and Greg Storrar,
MONITOR,
Movement composite model image,
2019
MONITOR is a cross between a vehicle
and a design studio. It is named as such
because of its hydraulically activated
movement patterns.

MONITOR
THE REBIS WALL

Instead of conceiving of relations between fixed identities, between entities or things that are only externally bound, the in-between is the only space of movement, of development or becoming. [...] This in-between is the very site for the contestation of the many binaries and dualisms that dominate Western knowledge [...].
— Elizabeth Grosz, *Architecture from the Outside: Essays on Virtual and Real Space*, 2006[1]

MONITOR is a sculptural, inhabitable, mobile, optical machine. It is an ongoing creative collaboration between artist James Capper (who makes operational sculptures consisting of mechanical parts and hydraulics), designer and researcher Thomas Pearce and architect Greg Storrar. It utilises robotic fabrication technologies in its construction. The inside of the 8-metre-long (26-foot) structure can be used as a moving creative studio framing all manner of changing views of its shifting situational contexts.

MONITOR was first conceived in 2018 for the residency programme of the Kyiv-based art gallery IZOLYATSIA. Having become nomadic since the first Russo-Ukrainian War in 2014, the gallery has found a temporary home in a warehouse building on the shipyard of Kyiv's harbour. This harbour is to be the base station for MONITOR, from where it can voyage up and down the Dnipro River, its movement predicated on actuating legs and a hinged torso that give it a lizard's gait. As its creators explain, 'It will house resident artists on expeditions into unfamiliar territories, aiming to explore, observe, and broadcast in times of increasingly wavering truth. Like an architectural Swiss-army knife, its thick walls are designed to contain a tight nesting of domestic and artistic functions needed for its artistic expeditions.'[2]

Multivalent Tropes

As a student, it was often suggested to me and my peers that all design choices made for architectural elements should be multivalent. That is to say, all architectural decisions should have more than one function and purpose. An example could be a column that is not just a structural member but also designed to be semiotic, voluptuously tactile and to support a light and a seat simultaneously, as well as its rhythm with other columns choreographing the emerging spatial episodes of a building. Looking at MONITOR INCREMENTAL WALL, the first full-scale prototype for MONITOR, reminds me of these pedagogic conversations.

The prototype consists of two panelled skins fabricated using robotic incremental sheet forming. The wall's unusual topology is created by the objects it stows as well as by the imprints of bodies inhabiting the studio. 'The sculptural piece develops as a conversation between the two skins, between the artistic and the domestic and is marked by transformations, mistranslations and echoes passing through the wall.'[3]

Equally, and perhaps more strangely, the wall also reminds me of one of my favourite books: Francesco Colonna's *Hypnerotomachia Poliphili* (1499). Poliphilo, as the main protagonist is called, stumbles along in a search for his love, Polia (many things), fuelled by a lusty interest in all that is animate and inanimate. He is a desiring observer full of hunger for aesthetic and sensual experiences. Throughout the book he progresses from one set piece to another, each concerning one architectural trope. This book is full of amazing ideas and illustrations. I use it in my lectures to introduce a series of tropes that are ubiquitous throughout the history of art and architecture. Such tropes include the

James Capper, Thomas Pearce
and Greg Storrar,
MONITOR,
Contextual Collage,
Dnipro River,
2022
opposite: This simple collage illustrates MONITOR's ability to use the Dnipro River as a means to travel further afield from its dockland warehouse base in the Ukrainian capital, Kyiv.

James Capper, Thomas Pearce
and Greg Storrar,
MONITOR INCREMENTAL WALL
prototype,
2022
left: MONITOR INCREMENTAL WALL is the first full-scale prototype of MONITOR. The orange external skin has a reciprocal relationship with the internal skin: what happens inside is reflected outside as a series of bulges, and vice versa – here, a chair is stored. The chair's back can be seen as creating a fabric-like effect on its adjacent outer wall.

ideal temple, alchemy, Arcadia, enigma, machines, vectors, collage/assemblage and bodies, to name but a few.

From an alchemical standpoint, we might consider MONITOR as a contemporary Rebis. The Rebis represents the co-joining of opposites and their reconciliation in both the real and the spiritual world – offering balance united in hermaphroditic form. The wall has these qualities; a state of in-betweenness, both profound and mundane, constructed by both physical and virtual technologies – some analogue, others digital. It might not be stretching it too far to see MONITOR's overall design as a contemporary ideal temple cossetting the tools, semiotics and bodies – both actually and metaphysically – for expedient creative journeys in dangerous terrain.

For MONITOR its projected travels will be through a mechanistic Arcadia full of enigma and grandiose architectural presences. Its design is influenced by those very same mechanistic typologies – it is a creature of the docklands. 'The typology of the studio, addresses and draws attention to the challenges and opportunities of Kyiv's post-industrial landscapes and architecture […] The studio's design hybridises influences from its infrastructural environment; a scientific research station functionality and aesthetics, fabrication technologies born in the shipping industry and developed through new cutting-edge research at [the Bartlett School of Architecture,] University College London [UCL].'[4] Whilst inspired by the infrastructural grandness of its shifting context, MONITOR is an intense small space that enables monk-like quiet personal study, allowing for only one artistic or domestic activity such as drawing, writing, reading or sleeping to take place at any one time.

James Capper, Thomas Pearce and Greg Storrar,
MONITOR INCREMENTAL WALL prototype,
2022

below: This image of the internal skin shows the size of the prototype relative to the human scale. Every space is utilised and every surface contoured.

opposite: Detail of the internal skin. The prototype facilitates the stowage of a variety of disparate materials and implements, particularly in its foam insulation interstitial layer between skins.

Robotic Wall Fabrication

The furniture, fittings and materials needed for these activities fold, roll or rotate out of the container-like volume of MONITOR when active. When not in use, they are stored in its walls and traces of them bulge out of the structure's exterior elevations – an imprinted catalogue of the activities taking place on the inside. 'This is taken as an opportunity to elevate and stage these rituals, framing them in dialogue with their environment and the architecture.'[5]

The technique used to form the metal panels of the prototype wall is known as 'roboforming' (short for robotic single-point incremental sheet forming), which refers to the process of plastically deforming sheet metal materials by gradually pressing the metal into shape using an industrial robotic arm. The robotic arm holds a brass stylus that forms the metal (0.9mm steel and aluminium) by following tool paths digitally generated from 3D geometry. As opposed to traditional sheet metal forming, the process does not require dies, hence affording complete customisation of each individual panel with minimum material wastage.

The robotically formed steel skin was developed and fabricated by a research group led by Pearce at the robotics lab of the Bartlett School of Architecture, UCL. It is this facility to customise the metal in this way that enables the wall to present its hermaphrodite, Rebis-like adaptable form both internally and externally. The research group developed a series of tactics and conversations between architectural elements, found objects and the bodies that use them to start to articulate the wall's errant and vagrant geometries. 'Tools and pieces of furniture are 3D-scanned and impressed into the panelled skin so they fit snuggly when stowed into the wall. Similarly, bodily movements are inscribed into the geometry of the panels. These impressions also become legible through the wall.'[6] For example, an axe head pressed into the interior skin bulges outwards to offer a boot scraper on the outside. The team liken the wall to the amphibious MONITOR creature it is a part of, as the wall is also understood as an in-between being.

BODIES AND THEIR MOVEMENTS ARE COMPLICIT IN SOME OF THE WALL'S ARTICULATIONS AND METAMORPHOSE INTO OTHER FUNCTIONAL TOPOLOGIES

Spanner, Chair and Knee

The correspondence between the two skins of the wall is a conversation marked by transformations, mistranslations and echoes passing through it.

A series of objects found in Capper's studio have been selected to be taken on MONITOR's expedition and are incorporated into the design for MONITOR INCREMENTAL WALL. Among them are a large spanner and a chair. These are 3D-scanned and pressed, digitally, into the thick wall. As an impression on the outer skin, the chair's turned legs and backrest stiles appear almost textile-like. 'When in use, the chair nestles underneath a desk, which folds elegantly out of the wall, revealing drawing utensils stored in the wall's foam interstice. The desk's shape follows the contours of surrounding functions and objects, such as the angle-poise lamp strapped onto and pressed into the wall above.'[7] Equally, bodies and their movements are complicit in some of the wall's articulations and metamorphose into other functional topologies.

Underneath the desk, the swing of the artist's knee, taking a seat, is expressed in the aluminium skin's surface. The kneecap is traced by a secondary toolpath evoking its patella structure and becomes the first rung of the exterior boarding ladder. These and many other programmatic objects/bodies are actors in the constitution of the complex skin distortions of the wall.

The multivalency of this prototype wall blurs the conditions between the simple legislation of spatial boundaries and tools/furniture – an old architectural trick yet seldom done with such dexterity, augmented by digital technology and metal-kneading robots. It is not just that this fragment of MONITOR is created by instrumental digital processes but that it evokes a poetic presence in its fabricated multiplicity that straddles the line between the human and the machinic, between action and reaction and proactive personal creation – a catalyst for change that is not just individual but architectural and community centred.[8]

James Capper, Thomas Pearce and Greg Storrar,
MONITOR INCREMENTAL WALL prototype,
2022

External skin detail. The stepped 'knee' protrusion on the outside becomes steps and aids boarding of the MONITOR vessels and external access for maintenance.

Notes
1. Elizabeth Grosz, *Architecture from the Outside: Essays on Virtual and Real Space*, MIT Press (Cambridge, MA and London), 2006, pp 92–3.
2. James Capper, Thomas Pearce and Greg Storrar, MONITOR, unpublished pdf, 2021.
3. *Ibid*.
4. *Ibid*.
5. *Ibid*.
6. *Ibid*.
7. *Ibid*.
8. MONITOR INCREMENTAL WALL and the research into robotic incremental forming is led by Thomas Pearce at the Bartlett School of Architecture, UCL, assisted by Gary Edwards, Theo Tan and Cristina Garza. The project is made possible thanks to generous support from the Higher Education Innovation Fund, Izolyatsia (Kyiv), the Bartlett Architectural Project Fund, Tata Steel, B-made at UCL, and Hannah Barry Gallery (London). See Thomas Pearce and Gary Edwards, 'Remote Impressions: Roboformed Prototypes for a Nomadic Studio', *ACADIA [Association for Computer Aided Design in Architecture] 2020: Distributed Proximities*, 2021, pp 2–6, and Thomas Pearce, 'An Architecture of Parallax: Design Research Between Speculative Historiography and Experimental Fabrication,' PhD thesis, University College London, 2022.

Text © 2024 John Wiley & Sons Ltd. Images courtesy Capper/Pearce/Storrar, photographs by Greg Storrar

CONSTRUCTING CHANGE
THE IMPACT OF DIGITAL FABRICATION ON SUSTAINABILITY

Fabio Amicarelli is an Italian architect and PhD researcher at the Academy of Architecture in Mendrisio, Università della Svizzera Italiana (USI) in Switzerland, with a strong interest in sustainable building processes and digital fabrication. He received his Master of Science in Architecture from USI, and has gained professional experience at CCL Architects in Lugano and ROK, Rippmann Oesterle Knauss in Zurich, where he worked as specialist in computational design and fabrication. As part of his doctoral project 'Foldcast', he developed paper-based formworks to shape structurally optimised concrete elements.

Ana Anton is a postdoctoral researcher at the Chair for Digital Building Technologies at the Institute of Technology in Architecture, ETH Zurich, and is associated with the National Centre for Competence in Research (NCCR) Digital Fabrication research consortium, where she leads the research in 3D concrete printing. She received her doctorate from ETH Zurich in 2022, and her architectural degree, cum laude, from TU Delft in 2014. While her scientific research addresses complexity and emergence in architecture, her designs exploit materiality encoded for digital fabrication. Her doctoral thesis, 'Tectonics of Concrete Printed Architecture', focuses on robotic concrete extrusion processes for large-scale building components.

Inés Ariza is an architect and researcher within the NCCR Digital Fabrication consortium at Gramazio Kohler Research, ETH Zurich, and R&D Project Lead at MESH AG. With a broad background in construction, robotic fabrication, additive manufacturing, circular engineering and computational design, her focus in on how disciplinary architectural methods are confronted by new technologies. Her recent research investigates concepts and methods for computational detailing when using robots to build. She holds a Master of Science in Design and Computation from the Massachusetts Institute of Technology (MIT) and a Doctor of Science in Digital Fabrication from ETH Zurich.

Tobias Bonwetsch is co-founder of Rematter, a startup that is committed to creating circular, low-carbon and equitable buildings. He is an architect and undertook his PhD in architecture and robotics at ETH Zurich. He has been at the forefront of developing robotic automation processes for architecture and construction for more than 15 years, both in research at ETH Zurich, and industry with the ETH spinoff ROB Technologies.

Mario Carpo is Reyner Banham Professor of Architectural History and Theory at the Bartlett School of Architecture, University College London (UCL). His research and publications focus on the history of early modern architecture and on the theory and criticism of contemporary design and technology. His award-winning *Architecture in the Age of Printing* (MIT Press, 2001) has been translated into several languages. His books include *The Alphabet and the Algorithm* (2011); *The Second Digital Turn: Design Beyond Intelligence* (2017); and *Beyond Digital: Design and Automation at the End of Modernity* (2023), all published by MIT Press.

Sasha Cisar leads sustainability research and active ownership at radicant bank, Switzerland's first digital sustainability bank. A trained architect, he has conducted research on architecture and building systems as well as sustainable construction at ETH Zurich. Before joining radicant, he was a sustainability analyst for real-estate and sustainability manager at Bank J. Safra Sarasin.

Benjamin Dillenburger is Professor of Digital Building Technologies at the Institute of Technology in Architecture, ETH Zurich. His research focuses on the development of building technologies based on the close interplay of computational design methods, digital fabrication and new materials. His work has been presented at the Venice Architecture Biennale, London Design Week and Art Basel Miami. His projects include two full-scale 3D-printed rooms exhibited at the FRAC Centre, Orléans, and the permanent collection of Centre Pompidou Paris, as well as the Smart Slab at the DFAB House in Dübendorf, Switzerland.

Kathrin Dörfler is an Assistant Professor of Digital Fabrication at the Department of Architecture within the School of Engineering and Design at the Technical University of Munich. She holds a Master's degree in architecture from TU Vienna and a PhD from ETH Zurich. Her research interests lie at the intersection of fabrication-aware design, collaborative fabrication processes, on-site robotics, and additive manufacturing for construction.

CONTRIBUTORS

Jelle Feringa is an architecture and robotics specialist. He co-founded EZCT Architecture & Design Research in 2001. While developing his PhD thesis at TU Delft, he developed the technical underpinnings for Odico formwork robotics, the first publicly traded architectural robotics company which he co-founded in 2012. From 2017, as Chief Technology Officer of Aectual, he was responsible for the development and production of tailor-made building products at a mass scale. In the summer of 2021, he founded a new initiative, Terrestrial, focused on high-volume, 3D-printed, large earthen structures. On the basis of this technology, Terrestrial is developing a next-generation acoustic barrier.

Corentin Fivet is Professor of Architecture and Structural Design at the Ecole Polytechnique Fédérale de Lausanne (EPFL) in Switzerland where he develops computational methods, design processes and construction techniques to increase the circularity of building structures. Under his direction, the Structural Xploration Lab is a front-runner in the areas of structural optimisation and ground-breaking large-scale demonstrator projects with reclaimed waste. He holds a PhD from UCLouvain in Belgium, and undertook a two-year postdoctoral stay at MIT.

Robert J Flatt has been a Full Professor and Chair of Physical Chemistry of Building Materials at ETH Zurich since 2010. Prior to this he was Principal Scientist at Sika AG after his PhD research at EPFL and postdoctoral research at Princeton University, New Jersey. His research interests have included the intersection of chemistry and construction materials to promote innovation and sustainability in construction and the preservation of built cultural heritage.

Abel Gawel received his PhD in robotics and machine learning from ETH Zurich in 2018. After this, he was a postdoctoral fellow at the CRI Group at Nanyang Technological University, Singapore, and a senior scientist at the Autonomous Systems Lab at ETH Zurich, where he led the mobile construction robotics group. Since 2021 he has been a Principal Researcher in Computer Vision and Machine Learning with the Huawei European Research Center, Zurich. His research interests include simultaneous localisation and mapping, high-accuracy localisation, object recognition, and semantic scene understanding with applications in construction robotics, industrial inspection, and search-and-rescue robotics.

Fabio Gramazio is a Swiss architect, and Professor of Architecture and Digital Fabrication at ETH Zurich. He is also the co-founder of the architecture and design group Gramazio Kohler Research, which focuses on the integration of digital fabrication and robotic technologies into architectural design and construction. Gramazio has been a leading figure in the field of digital fabrication in architecture, and his research and practice have contributed significantly to advancing the use of robotic technologies in the building industry. He has lectured and exhibited his work worldwide, and has received numerous awards for his contributions to the field.

Norman Hack is an architect and computational design researcher. He holds Master's degrees with distinction in architecture from TU Vienna and the Architectural Association (AA) in London. His career includes coding architecture at Herzog & de Meuron's Digital Technologies Group. His interest in integrative digital design and fabrication methods led him to pursue a PhD with Gramazio Kohler Research, completing it with distinction at the NCCR in Digital Fabrication at ETH Zurich. From 2018 to 2021 he held a tenure track professorship at TU Braunschweig, becoming a full professor of digital construction in 2022.

Tobias Huber is a partner at ZPF Ingenieure in Basel, Switzerland, where he leads the development of the hybrid earth-timber floor slab. He studied civil engineering at the University of Stuttgart, and graduated in 2005 from the Institute for Lightweight Design and Construction (ILEK). Since 2019 he has taught structural design at the Institute of Architecture at the University of Applied Sciences and Arts Northwestern Switzerland (FHNW).

Matthias Kohler is a Swiss architect and Professor of Architecture and Digital Fabrication at ETH Zurich. He is also the co-founder of the architecture and design group Gramazio Kohler Research, which focuses on the integration of digital fabrication and robotic technologies into architectural design and construction. He has been a leading figure in the field of digital fabrication in architecture, and his research and practice have contributed significantly to advancing the use of robotic technologies in the building industry. He has lectured and exhibited his work worldwide, and has received numerous awards for his contributions to the field.

CONTRIBUTORS

Silke Langenberg is Full Professor for Construction Heritage and Preservation at ETH Zurich. Her professorship is associated to the Institute for Monument Preservation and Historic Building Research as well as to the Institute of Technology in Architecture. She studied architecture in Dortmund and Venice. At ETH Zurich she addresses theoretical and practical challenges in the inventory and preservation of monuments as well as younger building stocks. Since her engineering dissertation, her research has focused on the rationalisation of building processes as well as the development, repair and long-term preservation of serially, industrially and digitally manufactured constructions.

Daniela Mitterberger is an Assistant Professor of Architecture and Digital Fabrication at Princeton University. She is also co-founder and director of MAEID (Büro für Architektur und Transmediale Kunst), a multidisciplinary architecture practice based in Austria. Her work focuses on novel ways of designing and fabricating architecture using extended-reality systems, human-machine collaboration, construction robotics and embodied computation. She received her doctorate from ETH Zurich with her thesis on adaptive digital fabrication and human-machine collaboration in architecture. Previously, she worked as a senior researcher at ETH Zurich and was Co-lead of the Immersive Design Lab as part of the Design++ initiative.

Yamini Patankar is a conservation architect and PhD student at ETH Zurich. She completed her Master of Architecture (Conservation) at the School of Planning and Architecture, Bhopal. Her research is focused on fostering multidisciplinary collaboration by integrating modern technology and visualisation techniques in the preservation of built heritage and beyond.

Robin Rehm is an art historian and senior scientist at the Institute for Preservation and Construction History at ETH Zurich. He is head of research at the Chair of Construction Heritage and Preservation and co-director of the research project Architecture and Patent: The Buildings of the ETH Domain. He was previously a lecturer at the Institute of Art History at the University of Regensburg in Germany, and a senior researcher at ETH Zurich. His research activities concentrate on the theory and history of architecture as well as interiors and furniture of the 19th and 20th centuries.

Romana Rust is a computational architect and researcher and is Head of Research and Innovation at vyzn AG. She has a Master's degree in Architecture from TU Graz, and a Doctor of Science in Digital Fabrication from Gramazio Kohler Research, ETH Zurich. Her particular interest involves the development of intuitive and innovative computational methods that integrate multiple design objectives such as geometry, acoustics, materiality and fabrication, including computational form-finding, fabrication-aware design, performance-based generative design and human-machine interfaces. She is also a member of the core developer team of COMPAS, an open-source computational framework for research and collaboration in AEC.

Sarah M Schlachetzki is scientific collaborator at the Institute for Preservation and Construction History at ETH Zurich. She is co-director of the research project Architecture and Patent: The Buildings of the ETH Domain. From 2014 to 2022 she was a junior faculty member at the Department of Architectural History and Preservation at the University of Bern. She was also previously a visiting fellow at New York University, and a guest researcher at the Humboldt University of Berlin. Her research interests include the history of architectural modernism as well as architecture in (East) Central Europe from the 18th to the 21st century.

Neil Spiller is Editor of $\mathcal{D}$, and was previously Hawksmoor Chair of Architecture and Landscape and Deputy Pro Vice Chancellor at the University of Greenwich in London. Prior to this he was Vice Dean at the Bartlett School of Architecture, UCL. He has made an international reputation as an architect, designer, artist, teacher, writer and polemicist. He is the founding director of the Advanced Virtual and Technological Architecture Research (AVATAR) group, which continues to push the boundaries of architectural design and discourse in the face of the impact of 21st-century technologies. Its current preoccupations include augmented and mixed realities and other metamorphic technologies.

Timothy Wangler is a Senior Researcher in the Physical Chemistry of Building Materials group at ETH Zurich. Prior to this he was a postdoctoral researcher at the Empa institute in Dübendorf, Switzerland, and a PhD researcher at Princeton University, with an interest in construction materials, chemical engineering and materials science, from digital fabrication with concrete to the conservation of built cultural heritage.